R & D, Innovation and Industrial Structure

Contributions to Economics

Albrecht Ritschl
Prices and Production
- Elements of a System - Theoretic Perspective -
1989. 159 pp. Softcover DM 59,-
ISBN 3-7908-0429-0

Arnulf Grübler
The Rise and Fall of Infrastructures
- Dynamics of Evolution and Technological Change in Transport -
1990. 305 pp. Softcover DM 85,-
ISBN 3-7908-0479-7

Peter R. Haiss
Cultural Influences on Strategic Planning
1990. 188 pp. Softcover DM 65,-
ISBN 3-7908-0481-9

Manfred Kremer/Marion Weber (Eds.)
Transforming Economic Systems: The Case of Poland
1992. 179 pp. Softcover DM 69,-
ISBN 3-7908-91415-3

Marcel F. van Marion
Liberal Trade and Japan
1993. 298 pp. Softcover DM 90,-
ISBN 3-7908-0699-4

Michael Carlberg
Open Economy Dynamics
1993. 203 pp. Softcover DM 75,-
ISBN 3-7908-0708-7

Hans Schneeweiß/
Klaus F. Zimmermann (Eds.)
Studies in Applied Econometrics
1993. 238 pp. Softcover DM 85,-
ISBN 3-7908-0716-8

Gerhard Gehrig/Wladyslaw Welfe (Eds.)
Economies in Transition
1993. 292 pp. Softcover DM 90,-
ISBN 3-7908-0721-4

Alfred Franz/Carsten Stahmer (Eds)
Approaches to Environmental Accounting
1993. 542 pp. Softcover DM 178,-
ISBN 3-7908-0719-2

János Gács / Georg Winckler (Eds.)
International Trade and Restructuring in Eastern Europe
1994. 343 pp. Softcover DM 98,-
ISBN 3-7908-0759-1

Christoph M. Schneider
Research and Development Management: From the Soviet Union to Russia
1994. 253 pp. Softcover DM 85,-
ISBN 3-7908-0757-5

Valeria De Bonis
Stabilization Policy in an Exchange Rate Union
1994. 172 pp. Softcover DM 75,-
ISBN 3-7908-0789-3

Bernhard Böhm / Lionello F. Punzo (Eds.)
Economic Performance
1994. 323 pp. Softcover DM 98,-
ISBN 3-7908-0811-3

Karl Steininger
Trade and Environment
1995. 219 pp. Softcover DM 75,-
ISBN 3-7908-0814-8

Michael Reiter
The Dynamics of Business Cycles
1995. 215 pp. Softcover DM 75,-
ISBN 3-7908-0823-7

Michael Carlberg
Sustainability and Optimality of Public Debt
1995. 217 pp. Softcover DM 85,-
ISBN 3-7908-0834-2

Lars Olof Persson / Ulf Wiberg
Microregional Fragmentation
1995. 132 pp. Softcover DM 75,-
ISBN 3-7908-0855-5

Andreas Foerster
Institutional Analysis of Development Administration
1995. 234 pp. Softcover DM 90,-
ISBN 3-7908-0853-9

Ernesto Felli / Furio C. Rosati / Giovanni Tria (Eds.)
The Service Sector: Productivity and Growth
1995. 355 pp. Softcover DM 98,-
ISBN 3-7908-0875-X

Giuseppe Munda
Multicriteria Evaluation in a Fuzzy Environment
1995. 255 pp. Softcover DM 90,-
ISBN 3-7908-0892-X

Boris Maurer

R & D, Innovation and Industrial Structure

Essays on the Theory of Technological Competition

With 18 Figures

Physica-Verlag
A Springer-Verlag Company

Series Editors
Werner A. Müller
Peter Schuster

Author
Dr. Boris Maurer
Lehrstuhl für Volkswirtschaftslehre,
insb. Angewandte Mikroökonomie
Universität Mannheim
D-68131 Mannheim, Germany

ISBN-13: 978-3-7908-0900-8 e-ISBN-13: 978-3-642-95925-7
DOI: 10.1007/978-3-642-95925-7

Die Deutsche Bibliothek - CIP-Einheitsaufnahme
Maurer, Boris:
R & D, innovation and industrial structure : essays on the theory of technological competition / Boris Maurer. - Heidelberg : Physica-Verl., 1996
(Contributions to economics)
Zugl.: Mannheim, Univ., Diss.
ISBN-13: 978-3-7908-0900-8
NE: Maurer, Boris: R and D, innovation and industrial structure

88/2202-5 4 3 2 1 0 – Printed on acid-free paper

Acknowledgments

This dissertation was written while I was a student at the *Graduiertenkolleg (GK) "Allokation auf Finanz- und Gütermärkten"* at the Universität Mannheim, and during a stay at the I.D.E.I. at Université de Toulouse 1.

Volker Böhm, who chaired the GK for the first two years, gave us a pleasant start. Without the many invaluable discussions during lunch and the afternoon coffee break (which was an institution at his "Lehrstuhl") it would never have been such an excellent experience. During various stages, Konrad Stahl (my supervisor) and Siegfried Berninghaus provided hints and comments. Konrad Stahl has been a careful reader in the last stage of writing. His suggestions are reflected at several places in this dissertation.

Special credit goes to Wolfram Wissler. He became not only a good friend, but an inspiring partner in many discussions. Joachim Winter did a great job in reading the first draft of this dissertation, and provided the most valuable comments (not to forget that he made the coffee in the morning).

I have to thank Bruno Biais for the discussions that led to the model in Chapter 4, and Jai Pil Choi, who found a serious mistake in the first draft of this model. I met Paul de Bijl from Tilburg during my stay in Toulouse. The various disputes with him had strong influence on my own work. I benefited a lot from discussions with Jaques Cremer, Jean-Charles Rochet, Emmanuelle Auriol and Jean Tirole. Thanks also to Dan Kovenock (Purdue) and Patrick Bolton (Brussels) for valuable comments on the fourth and fifth chapter.

Last but not least I am indebted to my friends and to my wife, Lucinde Schleifer. Writing a dissertation without having them in the back would really have been a much more difficult undertaking.

Table of Contents

Chapter 1

Introduction

The common topic of this collection of studies is the interaction between *innovative activity* of firms and *industrial structure.* I call this interaction *technological competition.*

Firms invest into R&D in order to open up new or enlarge existing profit opportunities for the future. A successful R&D-project leads to an innovation. An innovation introduced into the market changes the competitive structure of the industry. At the same time the structure of the industry shapes the incentives to invest into R&D. What matters for these incentives is not so much the existing structure but the expected dynamic evolution of that industry which is again dependent on the innovative choice of firms.

Amongst other things, the dynamic of industry evolution is therefore rooted in the dynamics of ongoing innovative activity. Of course, this is not always the whole story. There are (more or less) exogenous factors, like knowledge spillovers from other sectors of the economy, technological breakthroughs in basic research that directly influence the state of competition in an industry by providing additional profit opportunities, etc. The same is true for exogenous changes in upstream markets or demand conditions.

My main interest here is not primarily to understand these exogenous forces, but to develop a theory of how the process of firms' innovative activity is shaped by competition and in turn shapes future competition between firms in an industry. The latter point has been emphasized in many theoretical and empirical studies since the work of Phillips [1966]. He argued that it might well be the case that the causal link between industry structure and innovation does not lead from structure to innovation, but that innovation is the major force shaping industry structure. His study of the aircraft industry (Phillips [1971])gave strong support to this hypothesis.

The interdependence between industry structure and innovative activity is one of the most fascinating and exciting fields in Industrial Organization. This is especially so since it challenges the traditional view on market processes which concentrates on the analysis of static efficiency in current production.[1]

Plan of the Book

Technological competition involves not only activities taken by firms to develop new processes and products. At the same time the whole range of exogenous institutions and strategic actions that influence innovative success come into place and should be relevant in a thorough analysis of innovative activity and its interdependence with market structure.

Three major topics are addressed:

- How does the division of profits between innovating firms in the post-innovation market influence the incentives to innovate? – And how can the division of profits be determined? (Chapter 3)
- Do firms have incentives to hold back information about the success of a research project? (Chapter 4)
- How does the liquidity of firms shape the incentives to innovate? (Chapter 5)

In all three models I assume that there is post-innovation competition in the industry. Legal protection of rents from innovation is imperfect or not available and therefore the appropriation of returns depends not only on being the first to innovate. In Chapter 3, the division of profits is dependent on *quality choices*, in Chapter 4 on the timing of information revelation and market entry, while in the last chapter (Chapter 5) on the strategic reaction of firms to liquidity constraints.

To give an introduction to the field, a review of the (game-theoretical)literature on innovation and market structure is provided in Chapter 2.

[1]Schumpeter [1942] forcefully argued in this direction: "Perfect competition is inferior, and has no title to being set up as a model of ideal efficiency" (p.106).

Chapter 2: Review of the Literature

How does industry structure change in response to innovation? The bulk of theoretical work ignores this question. Instead, in most studies it is assumed that the post-innovation industry structure is given in advance, and the incentives of firms to survive the step from the existing structure to the new one is analyzed. The most prominent examples are *patent races.*[2]

Since innovations open up new profit opportunities it might be a good approximation to assume that the first successful innovator gains a comparative advantage over those that are second, and that this advantage can be measured by some fixed monopoly rent. Under such a simplifying assumption it is natural to ask, how much a firm is willing to pay to be the lucky one that is awarded the monopoly rent. *Patent races serve as a tool to analyze the question whether firms over- or underinvest under competition compared to some benchmark (i.e. the solution to the problem of a social planner).*– The answers given in the literature are discussed in Chapter 2.2.

Karl Marx developed a theory in which the evolution of the capitalist system leads to self destruction because of capital accumulation in the hand of a few capitalists and the progressive expropriation of the masses. A system cannot survive when all resources end up being controlled by only a small group of firms or individuals. This basic story about the dynamics of capitalism as a self-defeating system was challenged by the younger Schumpeter [1912].He argued that there is a continuous challenge of established firms by creative entrepreneurs developing new ideas and innovations. These *perennial gales of creative destruction* provide structural changes inhibiting a development like that described by Marx. Small and medium sized firms emerge, find profit opportunities, grow and will later themselves be challenged by new firms and entrepreneurs.

This does not claim that innovation theory is the hot and decisive topic in the struggle between those two great ideas about the course of history, but essentially it boils down to the question whether there is a tendency towards increasing dominance and concentration in capitalist industrial structures.

[2]In a narrow definition, a *patent race* implies a winner-takes-all assumption. In most of the literature the term is used in a broader definition that includes almost all game theoretical models of technological competition. I prefer the taxonomy introduced by Dasgupta [1986]: A *race* is a game where it pays of to be the first to complete a task.

The basic question inherent in this debate is whether established firms have comparative advantages over new firms in developing and introducing innovations. This was not the basic problem of Marx, but the senior Schumpeter himself drew a rather pessimistic picture of the future of capitalism due to these comparative advantages. He feared the superiority of big, established and multinational conglomerates, being more efficient in utilizing and producing new ideas and innovations.

To clarify the discussion, a differentiation into two basic questions is helpful. The first question is whether established firms or firms in concentrated industries have comparative advantages, the second, whether these firms have the incentives to use them.[3] The question concerning the incentives to innovate has been the motivation for a number of researchers to build models analyzing the asymmetric situation between established firms in an industry and potential entrants in a (more or less) dynamic setting.

I review the literature on this topic in Chapter 2.3. Three methodologically different approaches can be distinguished. The first approach is to analyze a sequence of drastic innovations by using *repeated patent races.* In the second approach industry structure changes gradually through small improvements and changes in production technology, or by the introduction of variants of new products. I call these studies *analyses of optimal adaptation.* In the last approach the research process itself is modeled as a *multi-stage game.*

A common force driving the results in most of these models is the strategic significance of innovative activity for market competition; in particular whether this activity results in *strategic substitutes* or *strategic complements* plays its role in almost all results.

Beside this strategic feature, the marginal expected return on investment when this is a continuous variable determines the extent to which firms are willing to invest into R&D. But not only the *size of marginal profits* is important; also the division of profits in the post-innovation market. It is therefore natural to extent the analysis to *sequential innovative entry* into a market, or to allow for the *possibility of imitation.* The existing literature on this point is reviewed in Chapter 2.4.

[3] A theoretical explanation of why established firms have comparative advantages, (namely, because of capital market imperfections) is given in Chapter 5.

Chapter 3: Sequential Innovative Entry

In most industries legal protection is not sufficient to guarantee post-innovation monopoly rents to a first innovator. Leadership nevertheless may be important to appropriate the returns from R&D, i.e. through the creation of a captured market or the possibility to deter future entry. Beside these arguments leadership creates disadvantages. Followers may learn from first movers without incurring own costs. In addition, if the first mover is committed to his current *technological standard* (or *product quality*) at least for some time, followers may improve upon the standard chosen by the first mover and thereby gain a larger profit share (or share of the market).

Starting out from that reasoning, the division of profits in the post-innovation market is a function of the technological standards adopted by different firms. Accordingly, in Chapter 3 I develop a framework where firms choose *technological standards.* The formal tool is a search-theoretical model or *stopping game.*

To allow for the analysis of the impact of quality choices of early and late movers a dynamic model is chosen where firms may enter the market at each point in time with their current technology. Once in the market, firms are committed to their current technology or quality. The currently available technology is the result of past research of the firm. Research is modeled as a stochastic process. Firms set up research projects of uncertain duration and outcome. The result may either be marketed, or a firm may decide to start another attempt in order to achieve a higher standard or quality. A firm that continues research may be preempted by a rival.

In the chosen framework, preemption does not necessarily mean that profits are spoiled for the preempted firm, since it can redesign its research strategy and enter later with a more advanced standard or higher quality.

The main results are:

- The division of expected profits between early and late innovators play a decisive role for the incentives to innovate and the underlying strategic structure of the game.
- The division of profits arises endogenously. There may be a multitude of equilibria with both first-mover advantages and second-mover advantages.

- A game with first-mover advantages and multiple equilibria is a game of coordination i.e., equilibria are symmetric and profits rise with the strategies played in equilibrium.
- It may be that equilibria with both first-mover advantages and second-mover advantages coexist. In this case entry tends to be early. The first innovator chooses an inferior quality, and the expected profits of a later entrant are higher.

Chapter 4: Strategic Use of Delay and Secrecy

In Chapter 3 the division of profits in the post-innovation market was dependent on the technological standards or quality supplied by first- and late movers. In turn, the incentives to provide high quality are determined by the emerging expected division of profits in equilibrium.

In Chapter 4 I return to the analysis of strategic behavior under the assumption that the result (i.e. the technological level of the innovation) is exogenously given. The firm that completes its research project has an innovation that it may bring to the market. Patent protection is insufficient to erect barriers to entry for subsequent firms. A second firm may enter at a later point in time. Leadtime is often considered to be a good means to appropriate returns from R&D, mainly because of learning effects or captive consumers. First entry may also create the possibility to use incumbency for entry deterrence.

In the approach taken in Chapter 4 I abstract from these factors to ask if there may also be incentives to defer the introduction of an innovation into the market. There are mainly two reasons to render profitable the delay of an innovation. First, *replacement effects.* A current incumbent in a market has less incentives to introduce a new product that spoils his current market. The literature on these effects is reviewed in Chapter 2.3. Second, *informational spillovers.* The first entrant into a market provides two types of information to potential competitors. The first type of information is about the technology or about market conditions. This is *data* relevant for own innovational efforts and construction. The second type of information is simply that *market success indicates that there is a profitable solution to a certain kind of problem.*

I concentrate on this last sort of informational spillovers. I use a variant of the *stopping game* in Chapter 3, with two modifications. First, the technological standard or quality of the innovation is exogenously given. Second, an additional source of uncertainty is introduced into the model. To understand the necessity for the latter modification, think about a situation in which the information "that a problem is solvable" is indeed of positive value. This happens to be true whenever one is not completely sure that the approach taken to solve a problem necessarily leads to any solution at all. I therefore introduce uncertainty about the potential productivity of the research approach taken in an industry. This uncertainty is (at least partially) resolved whenever one firm in an industry is successful and the success is made publicly known (i.e. by using it).

The main results are:

- Firms become increasingly pessimistic about the prospects of research in case of no observed success.
- "Early" success leads to direct introduction of the innovation. The first firm that succeeds enjoys a temporary monopoly position until a rival enters.
- "Late" innovations are kept secret over a certain delay period until potential rivals got sufficiently pessimistic to quit the race.

Chapter 5: Innovation under Financial Constraints

As in most of the literature, also the models of Chapter 3 and 4 ignore the source of finance for the R&D expenditures of the firms. Firms normally finance their R&D-projects out of retained profits and earnings. Especially for young and expanding firms nonexistent earnings pose a great problem. They have to rely on outside funding to get the chance to conduct their projects.

At the same time the funding of risky projects bears problems of asymmetric information between firms and outside claimholders. To insure against the risks involved in this principal-agent relationship rational lenders prefer *standard debt contracts* that provide short term debt with renewal conditioned on repayment. Financial constraints therefore may lead to a *weakened standing* of firms that are credit rationed.[4]

[4] A short digression into optimal contract theory is provided in Chapter 5.2. For a more complete review of the literature see Bhattacharya and Thakor [1993], Dowd [1992] (debt contracts) or v. Damme [1994] (banking).

Established firms with a long purse may have an incentive to use their longer breath for predatory actions to drive rivals out of the market. In Chapter 5 I present a two-period model with this feature. I show that a *standard debt contract* between an outside claimholder and a firm is an optimal contract. Given that a firm and its bank agree on such an optimal contract, the probability of finance in the future can be adversely affected by other firms in the industry in a dynamic setting. Unleveraged firms may choose to exert higher innovative effort in the presence of leveraged rivals for purely predatory reasons.

The main results are:

- The optimal contract between a firm and an outside investor that cannot directly be based on observable variables resembles a standard debt contract.
- In a multiperiod framework this contract provides insufficient finance compared to the first best. A firm that is unsuccessful in the first period may not always find finance even for an *ex ante* profitable investment in later periods.
- A leveraged firm can be adversely affected by increased rivals' innovative effort in the first period. This predatory behavior of rivals worsens the inefficiency of the contract.
- Outside finance may lead to either increased or decreased innovative effort of the leveraged firm, dependent on the basic strategic structure of the underlying game.
- An optimal contract that is not vulnerable to rival attack is likely to fail renegotiation proofness.

Chapter 2

A Review of the Literature on Technological Competition

2.1 Introduction

The purpose of this chapter is twofold. *First*, to provide an introduction into the *game-theoretical* literature on *technological competition*. *Second*, the review of the literature serves as a starting point and motivation for my own work. Each section concludes with a short discussion in which open questions are addressed that will be discussed in later chapters.

Since I am interested in a theory of competition, the interdependency of the actions taken by firms should be at the heart of the analysis. The older decision-theoretical approach is therefore inappropriate for the purpose.[1] As a theory of strategic interaction game theory evolved to be a unifying tool in almost all areas of theoretical industrial economics.

In the following I give a short plan of the chapter and the most important topics and results. The main topics are:

- Effects of *innovation competition* on the investment decisions of firms. These effects are analyzed in *one-shot models of technological competition.*

[1] For this literature see the seminal work of Kamien and Schwartz [1971, 1972, 1976], or their survey-monograph [1982].

- *Increasing dominance* and the question whether *success breeds success.* To analyze this question, one has to turn to dynamic *multi-stage models of technological competition.*

- Effects of *market competition* on the *investment decisions* of firms, as well as on the *timing of market entry* with a completed innovation. Models with *imperfect appropriability* of innovation returns and *sequential market entry* are used for this purpose.

One-Shot Technological Competition

The game-theoretical treatment of strategic investment into R&D started with the work of Dasgupta and Stiglitz [1980a], Loury [1979], and Lee and Wilde [1980]. All three papers concentrated on the question whether competition for a innovation of known profitability leads to over- or underinvestment into R&D (compared to the benchmark model of a social planner conducting research). Whether there is over-investment or underinvestment depends on how research is contracted. Contractual commitment of resources (i.e. subcontracting research) leads to overinvestment into R&D, while keeping discretion over R&D-expenditures (i.e. in-house research) leads to underinvestment. A formal exposition is found in Section 2.2.1.

Recently, Lippman and Mamer [1993] produced a model in which the profitability of an innovation is endogenously determined. Increased competition leads to innovations that are less profitable and of a lower technological standard (or product quality in case of product innovation). Their model is presented in Section 2.2.2.

A natural extension of the basic patent races in Section 2.2.1 is to include post-innovation competition into a two-stage model of innovation. Two approaches are discussed: The deterministic model of Dasgupta and Stiglitz [1980b], which produces a linear relationship between concentration and the R&D/sales-ratio, as well as the stochastic search model of Reinganum [1983a] that relates expected cost reduction through innovation to competition (and is therefore similar to the model of Lippman and Mamer [1993]). They are further discussed in Section 2.2.3.

Dynamic and Multi-Stage Technological Competition

The dynamic games of technological competition concentrate on the strategic use of innovative activity to achieve a dominant position in a market or to deter entry into a market.

Three methodologically distinct approaches are discussed in the literature. *First*, models with a sequence of *drastic innovations*. The standard reference here is Reinganum [1985]. The main question addressed is whether current monopolists stay dominant in a dynamic setting, or if there is some pattern of alternating incumbency to be expected. Two alternative explanations are offered: Persistence of monopoly is to be expected in environments without technological uncertainty (Reinganum [1983b, 1984]), and if the innovations are "less drastic" (Gilbert and Newberry [1982], and Vickers [1986]). Uncertainty favors the *replacement effect*, while less drastic innovations give rise to a *joint profit effect*.[2] The *replacement effect* explains why a current incumbent has less incentives to invest into innovation (namely, because she replaces herself in case of success), while the *joint profit effect* works in the other direction: Since monopoly profits are frequently higher than joint duopoly profits, a current monopolist can outbid a potential entrant in an auction for the exclusive right to use an innovation by bidding just above the expected duopoly profits of the potential entrant. Given that the innovation is not drastic, her profits are then still higher than from accommodating entry. The second explanation of the persistence of monopoly is given by Lippman and Mamer [1992]. The division of profits between winners and losers of innovation competition and the non-recoverability of R&D-expenditures is responsible for the persistence in there framework. See Section 2.3.1 for details.

Second, models with *incremental innovations*. These models provide a framework for the debate on whether there is *increasing dominance* in a market. There may be a reason for increasing dominance arising from the fact that a firm with a high market share extracts a higher additional profit from a marginal cost reduction than smaller competitors. Nevertheless, in the two models (Völker [1990] and Budd, Harris and Vickers [1993]) presented in Section 2.3.2 leading firms are likely to slow down (or cease) their innovational activity when the lead is "large".

Third, models of multi-stage research processes. The basic structure of these models is similar to the one-shot patent races in Section 2.2.1. The first firm that completes all stages of a research program is awarded a perfect patent of fixed value to the firm. In addition, firms receive information about their relative position to their rivals while they are researching. Firms may abandon research at any stage. The analysis concentrates on the question whether firms drop out without finishing the

[2] See also Gilbert and Newberry [1984] and Katz and Shapiro [1987].

task when far behind (and which effects this has on the leading firm), or if there will be vigorous competition and *rent dissipation.* In Section 2.3.3 I discuss the models of Fudenberg *ea.* [1983], Harris and Vickers [1987], Grossman and Shapiro [1987] as well as Lippman and Mc Cardle [1987]. The actually observable pattern of competition depends on the informational structure of the game and on the possibility of *catching up*, or *leap-frogging.*

Sequential Entry

There is strong empirical evidence that technological knowledge of innovators leaks out to rivals, and often innovation does not lead to *ex-post* monopoly. In addition, intellectual property rights regularly are insufficient to deter subsequent entry into a market.

The last section (Section 2.4) provides a theoretical analysis of phenomena which may be observed if sequential choice of technology from a given set is allowed for. McLean and Riordan [1989] analyze the behavior of first adopters that can choose from a menu of technologies. They may have incentives to choose technologies that *raise rival's costs*[3] and delegate the adoption of entry-deterrent technologies to later adopters. In Dasgupta [1986, 1988] the theoretical implications of first- and second-mover advantages are discussed in a static model. In the next chapter I show that there is a close link between those two ingredients, namely that relatively *high first-mover profits* result in *games with strategic complementarities*, while relatively high *second-mover profits* are closely related to the notion of *strategic substitutes*, where the strategies are the quality levels of the innovations that are finally adopted. Therefore this section serves as a link between this chapter and the model presented in Chapter 3.

In my review of the literature I use a unified notation and framework whenever possible. This allows to highlight the structural links between the models and the driving forces behind the different results. I hope that this attempt serves to reduce the complexity and variety of the different approaches used in the literature. Since the game-theoretical toolbox provides a framework for the analysis of the *incentives* to behave in a certain way in models of *strategic interaction*, it is often helpful to classify the *strategic structure* of the underlying games. I therefore repeatedly use the notion of *games with strategic complements* and *games with strategic substitutes.*

[3]See e.g. S. Martin [1993], p.69 on this point

Readers not familiar with this notion are referred to Bulow *et.al.* [1985].[4]

2.2 One-Shot Technological Competition

2.2.1 The Model with Perfect Patent Protection

The first game-theoretical treatments of innovation under competition have been introduced by Loury [1979], Dasgupta and Stiglitz [1980a], and Lee and Wilde [1980]. The underlying structure of the innovation process is identical in all three models. They analyze the question whether firms over- or underinvest in the case of innovation competition under free entry compared to the solution of the problem of a monopolist (or social planner). It turns out that the decisive factor is how research is contracted upon.

The Model

Consider n potential firms. Each firm can undertake a risky research project. It is assumed that research is contracted upon between each individual firm and a research unit inside or outside this firm. The time of successful completion of the innovation is a random variable which is influenced by the firm's research expenditures. The first firm to complete the project earns a flow profit V, the others earn nothing.

Two different contractual designs between a firm and a research unit are analyzed. In the first version, all expenditures are contracted upon at date 0. Thus, the firm purchases a random variable $\tau(x_i)$, where x_i are the R&D-expenditures and τ the random completion time. In the second version there is a set-up cost F to be payed at date zero, and a flow cost of y_i until the project is either completed or abandonned.

The first case resembles a situation where firms decide to *subcontract research.* The volume of the contract is specified *ex ante*, as well as the desired result. This presupposes that the necessary research expenditures can be correctly assessed by the firm in advance (which is a rather strong assumption in the case of innovative activity). The second case corresponds more to the case of *in-house research* where the firm retains discretion about the project.

[4]See also Fudenberg and Tirole [1984] for a similar taxonomy.

First, I treat the case with contractual commitment.[5] The completion time $\tau(x_i)$ is assumed to be an exponentially distributed random variable with parameter $h(x_i)$, and $h(x) \in C^2$ (twice continuously differentiable); $h(0) = 0, \lim_{x\to\infty} h'(x) = 0$, and

$$h''(x)\begin{cases} > 0 & , \quad if\ x < \bar{x} \\ = 0 & , \quad if\ x = \bar{x} \\ < 0 & , \quad if\ x > \bar{x}. \end{cases}$$

Interpretation: The R&D-technology exhibits initially increasing returns to investment up to a threshold $\bar{x}$, and decreasing returns thereafter. The justification for this assumption is that investment at a very low scale is not optimal, i.e. a certain critical mass of equipment and staff is needed to tackle with the task efficiently.[6] All firms are assumed to be identical, and only symmetric equilibria are analyzed. The model incorporates two sources of uncertainty. First there is *technological uncertainty* about the time of own success, and second there is *market uncertainty* about the first success of a rival firm represented by $\hat{\tau}_i = \min_{j\neq i} \tau(x_j)$, which is the minimum completion time of the rivals of i. Thus we have

$$\begin{aligned} P[\tau(x_i) \le t] &= 1 - e^{-h(x_i)t} \qquad \text{and} \\ P[\hat{\tau}(x_i) \le t] &= 1 - e^{-\Sigma_{j\neq i} h(x_j)t} := 1 - e^{-a_i t} \end{aligned}$$

Since firm i earns flow profits V if $\tau(x_i) < \hat{\tau}_i$, the innovation has a present value upon the first completion date of V/r where r is the discount rate. Using the appropriate discount factor, the problem of the firm becomes

$$\max_{x_i} \Pi(a_i, r, V; x_i) = \frac{V}{r}\left(\frac{h(x_i)}{a_i + r + h(x_i)}\right) - x_i \tag{2.1}$$

[5]This formulation is due to Loury [1979].

[6]This is the "minimum efficient scale" argument. For the empirical relevance see e.g. Markham [1965]. To say the least, this position is controversial. The counter-position is formulated by Jewkes, Sawers and Stillerman [1958]:

> "Even where inventions have arisen in research laboratories of firms, the team responsible for it seems often to have been quite small. It is usually found that one outstanding figure has been surrounded by, and has stimulated, a few devoted collegues in an intimate relation with his manner of thought and speculations" (p.88).

However, the truth should be somewhere in the middle: There is a minimum scale of accumulated human capital (and simply: hours of work) necessary to be successful in a reasonable time.

The necessary and sufficient conditions for optimal investment are given by

$$\frac{V}{r} = \frac{(a_i + r + h(x_i))^2}{h'(x_i)(a_i + r)}, \tag{2.2}$$

$$\frac{V}{r} \leq \frac{2h'(x_i)(a_i + r + h(x_i))}{h''(x_i)(a_i + r)}. \tag{2.3}$$

The equilibrium investment by each firm is given by

$$x^* = x((n-1)h(x^*), r, V, n). \tag{2.4}$$

It can be shown that the solution is unique. To determine the number of firms willing to invest, a zero expected profit entry constraint is added.[7] The expected profit of each firm is a function of n, and is given by

$$\Pi(x^*, n) = \frac{h(x^*)}{h'(x^*)} \left(\frac{nh(x^*) + r}{(n-1)h(x^*) + r} \right) - x^* \tag{2.5}$$

The equilibrium number of firms solves

$$\Pi(x^*, n(x^*)) = 0 \tag{2.6}$$

Investment at an efficient scale requests the equality between average and marginal instantaneous success probability (This relation is obtained by solving the problem for a monopolist or social planner conducting the optimum number of parallel projects.). Actually, from the conditions for the optimal solution one has

$$\frac{h(x^*)}{x^*} = h'(x^*) \frac{a + r}{a + h(x^*) + r} \leq h'(x^*), \qquad h(x^*) > 0. \tag{2.7}$$

Thus, each firm builds excess capacity compared with the efficient outcome.

In the second case,[8] where there are no contractual commitments over the investment plan, a firm chooses flow expenditures y_i to solve[9]

$$\max_{y_i} \Pi(V, r, a_i; F, y_i) = \frac{\frac{V}{r} h(y_i) - y_i}{a_i + h(y_i) + r} - F \tag{2.8}$$

[7]For the details of the model with long-run competition under free entry, see Dixit and Stiglitz [1977].

[8]The no-commitment case was first treated by Lee and Wilde [1980].

[9]The rate of R&D-expenditures is fixed once and for all at a constant level for simplicity. For the time profile of optimal investment paths for risky projects, see Kamien and Schwartz [1971] and Grossman and Shapiro [1986].

This leads to the first order condition

$$\frac{V}{r} = \frac{a_i + r + h(y_i) - y_i h'(y_i)}{(a_i + r)h'(y_i)}. \tag{2.9}$$

In the equilibrium with free entry, each firm i chooses an investment flow which leads to a marginal instantaneous probability of success given by

$$\frac{h(y_i)}{y_i + (a_i + r)F} = h'(y_i) < \frac{h(y_i)}{y_i}. \tag{2.10}$$

This time each firm underinvests in R&D.

To sum up the results of the two basic versions of the one-shot patent race it can be stated that the investment into R&D under rivalry is always inefficient. Whether there is over- or underinvestment depends on the design of the research contract. Contractually committed expenditures correspond to expenditures being *strategic substitutes*, while R&D flow-expenditures are *strategic complements* in the case of no commitment.[10]

A further result is that the firms in both specifications invest more in aggregate terms than a monopolist (or social planner) undertaking parallel research projects would invest.

2.2.2 Endogenous Technology Choice

While the models in the last section had a "winner-takes-all" assumption together with a fixed price, the model of Lippman and Mamer [1993] endogenizes this price. Firms are free to enter a market at any point in time. Once the first firm enters it is

[10]To see that expenditures are strategic substitutes under commitment, note that (totally differentiate equation 2.2)

$$\frac{dx_i}{da_i} = -\frac{h'(x_i)(a_i + r + h(x_i) - 2h(x_i))}{2(a_i + r)((h'(x_i))^2 - (a_i + r + h(x_i))^2 h''(x_i))},$$

which is negative in case of excess capacity and $n \geq 2$. In the case of no commitment it holds that (totally differentiate equation 2.9)

$$\frac{dy_i}{da_i} = \frac{h'(y_i)(h(y_i) - y_i h'(y_i))}{h''(y_i))(a_i + r)(a_i + r + h(y_i))},$$

which is positive in case of underinvestment.

awarded a perfect patent for its current best technology (or product quality). The technological or quality standard stochastically increases over time. The resulting game is a *stopping game.*[11]

The Single-Firm Case

Consider a monopolist investing in R&D at a constant rate c in order to develop a new product. Research is organized as follows. There is a sequence of (identical) research projects. Each project eventually results in a readily developed new product prototype of a certain quality.

In the present setting the *quality* of a product is identified with its expected profitability. More advanced products guarantee higher future profit streams. Thus, the quality is measurable by a unidimensional parameter, i.e. expected profits. Let this parameter be denoted by γ. The outcome of a research project is thus an innovation with some profitability index γ, and I assume that γ is distributed according to some absolute continous c.d.f. $F(\gamma)$ which possesses a density $f(\gamma)$ on the positive real numbers.

The arrival time of an innovation is random. It follows a poisson arrival process with parameter λ. After each realization the firm decides whether to continue or to abandon research and enter the market with the currently best prototype. The problem of the firm is to choose an *optimal stopping-time* τ for the research process. A stopping time in this framework is defined as a random variable with the property that it is the minimum time where the state variable γ reaches some stopping region Γ.[12] It can thus be written as

$$\tau := \min\{t \in \mathbb{R}_+ | \gamma_t \in \Gamma\}. \tag{2.11}$$

This is equivalent to the choice of a threshold quality $\bar{\gamma}$ with the property that $\bar{\gamma}$ is the minimum quality in the stopping region Γ.

[11]The model of Lippman and Mamer [1993] is presented in greater detail, since the same framework is used later in Chapter 3.

[12]In more general settings this stopping region may be time-dependent as well. However, Lippman and Mamer restricted the analysis to that of stationary strategies. This seems reasonable in a stationary environment.

The arrival time $T_{\bar{\gamma}}$ of an innovation with a quality of at least $\bar{\gamma}$ is exponentially distributed with parameter $\lambda(1 - F(\bar{\gamma}))$. Let the discount rate r be constant over time and positive. The appropriate discount factor to calculate the present value of the profit stream is given by

$$\begin{aligned} E[e^{-rT_{\bar{\gamma}}}] &= \int_0^\infty e^{-rT_{\bar{\gamma}}} dF(T_{\bar{\gamma}}) \\ &= \int_0^\infty e^{-rt}\lambda(1 - F(\bar{\gamma}))e^{-\lambda(1-F(\bar{\gamma}))t}dt \\ &= \lambda(1 - F(\bar{\gamma}))\int_0^\infty e^{-(r+\lambda(1-F(\bar{\gamma})))t}dt \\ &= \frac{\lambda(1 - F(\bar{\gamma}))}{r + \lambda(1 - F(\bar{\gamma}))}. \end{aligned} \tag{2.12}$$

Therefore, the present value of R&D-expenditures is given by

$$\begin{aligned} -cE[\int_0^{T_{\bar{\gamma}}} e^{-rt}dt] &= -c\int_0^\infty \int_0^{T_{\bar{\gamma}}} e^{-r\tau}d\tau dF(T_{\bar{\gamma}}) \\ &= -\frac{c}{r}\int_0^\infty (1 - e^{-rT_{\bar{\gamma}}})dF(T_{\bar{\gamma}}) \\ &= -\frac{c}{r}E[1 - e^{-rT_{\bar{\gamma}}}] \\ &= -\frac{c}{r}(1 - \frac{\lambda(1 - F(\bar{\gamma}))}{r + \lambda(1 - F(\bar{\gamma}))}) \\ &= \frac{-c}{r + \lambda(1 - F(\bar{\gamma}))} \end{aligned} \tag{2.13}$$

The value of the innovation upon entry is for simplicity identified with the realized quality γ. The firm thus chooses a threshold quality $\bar{\gamma}$ in order to maximize the present value profits, i.e.

$$\max_{\bar{\gamma}} \frac{-c + \lambda \int_{\bar{\gamma}}^{\infty} \gamma dF(\gamma)}{r + \lambda(1 - F(\bar{\gamma}))}. \tag{2.14}$$

The necessary (and also sufficient) condition to (2.14) is given by

$$r\bar{\gamma} = \lambda \int_{\bar{\gamma}}^{\infty} (\gamma - \bar{\gamma})dF(\gamma) - c \tag{2.15}$$

An Example

Let $F(\gamma) := (1 - e^{-\gamma})$. An optimal threshold $\bar{\gamma}$ for the monopolistic firm is a solution to (from equation (2.15))

$$r\bar{\gamma} = \lambda e^{-\bar{\gamma}} - c. \tag{2.16}$$

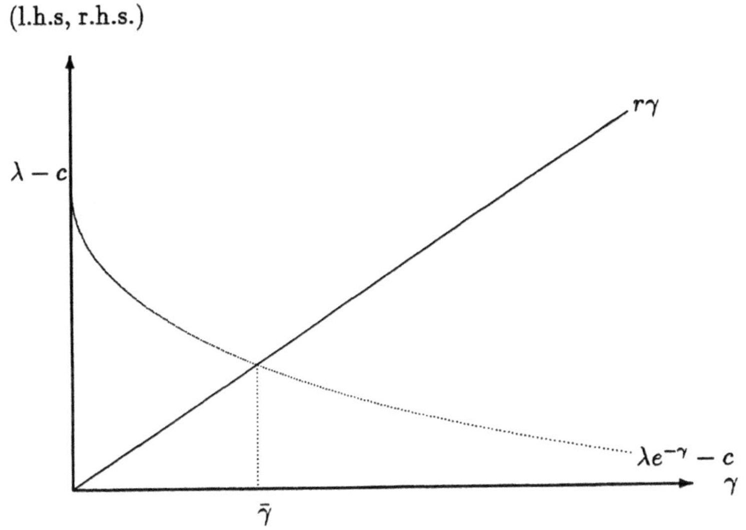

Figure 2.1: The Optimal Threshold for a Single Firm

The graphical solution is depicted in Figure 2.1. The optimal threshold is decreasing in the research cost c and in the discount rate r, while it is increasing in the technological "efficiency"-parameter λ.

Many Firms

Suppose now that there are n firms competing for the innovation. The first firm to enter is granted an eternal patent irrespectively of the realized quality. Let the profitability of the innovation again be identified with its quality.

Let me further assume that the game is with imperfect information, i.e. each firm observes only the evolution of its own quality over time.[13] Each firm chooses a threshold strategy $\bar{\gamma}^i$ that maximizes expected profits given the strategies of all other firms.[14]

[13] Under the assumption of independent successive draws which are distributed identically, and with the memoryless property of the process the informational structure of the game is irrelevant. This changes in the case where research is cumulative. I will return to this point later.

[14] Lippman and Mamer [1993] show that it implies no loss of generality to restrict the analysis to equilibria in stopping times which are representable by threshold levels. See Lemma 1 in their

Firms are characterized by their efficiency parameters λ_i, and their cost parameters c_i. I consider the problem of firm i. Given the threshold strategies of the other firms $(\bar{\gamma}_1, \ldots, \bar{\gamma}_{i-1}, \bar{\gamma}_{i+1}, \ldots, \bar{\gamma}_n)$, the time of the first rival innovation T_{-i} is exponentially distributed with parameter $\sum_{j \neq i} \lambda_j (1 - F(\bar{\gamma}^j))$. Let the time of success of firm i using its optimal threshold strategy be denoted T_i. Let $T := (T_i \wedge T_{-i})$ be the stopping time of the first entrant. The appropriate discount factor for the present value of the first marketed innovation is thus given by

$$E[e^{-rT}] = \frac{\sum_{i=1}^n \lambda_i (1 - F(\bar{\gamma}^i))}{r + \sum_{i=1}^n \lambda_i (1 - F(\bar{\gamma}^i))}. \tag{2.17}$$

The conditional expectation given that i preempts its rivals is

$$E[e^{-rT_i} | T_i < T_{-i}] = \frac{\lambda_i (1 - F(\bar{\gamma}^i))}{r + \sum_{i=1}^n \lambda_i (1 - F(\bar{\gamma}^i))}. \tag{2.18}$$

Firm i invests until it realizes an innovation of a higher quality than its threshold as long as it is not preempted by a rival. The expected expenditures of firm i until the first firm enters the market is therefore

$$\begin{aligned} - cE[\int_0^T e^{-rt} dt] &= -\frac{c_i}{r}(1 - \frac{\sum_{i=1}^n \lambda_i (1 - F(\bar{\gamma}^i))}{r + \sum_{i=1}^n \lambda_i (1 - F(\bar{\gamma}^i))} \\ &= \frac{-c_i}{r + \sum_{i=1}^n \lambda_i (1 - F(\bar{\gamma}^i))}. \end{aligned} \tag{2.19}$$

Putting pieces together leads to an expected profit of firm i given by

$$\frac{-c_i + \lambda_i \int_{\bar{\gamma}^i}^{\infty} \gamma^i dF(\gamma^i)}{r + \sum_{i=1}^n \lambda_i (1 - F(\bar{\gamma}^i))}. \tag{2.20}$$

The solution is derived using a dynamic-programming approach.[15] A threshold strategy $\bar{\gamma}^i$ is optimal if firm i is indifferent between stopping research at time t if $\gamma_t = \bar{\gamma}_t$, and waiting just a little bit longer until $t + dt$. Thus we have the following Bellman equation

$$\begin{aligned} \bar{\gamma}^i &= \max_{\gamma^i} \Big\{ - cdt + \lambda_i \int_{\gamma^i}^{\infty} \zeta dF(\zeta) e^{-rdt} dt \\ &\quad + (1 - (\sum_{j \neq i} \lambda_j (1 - F(\bar{\gamma}^j)) + \lambda_i (1 - F(\gamma^i))) dt) \gamma^i e^{-rdt} \Big\}. \end{aligned} \tag{2.21}$$

paper.

[15]This approach is used, because it is the standard approach for dynamic optimization problems. In the present formulation one can also directly use the first order condition derived from differentiating (2.20), since γ is not a function of time.

For a very short interval, $e^{-rdt} \approx 1 - rdt$. Ignoring terms of order (dt^2) or higher one arrives at

$$(r + \sum_i^n \lambda_i(1 - F(\bar{\gamma}^i)))\bar{\gamma}^i = -c_i + \lambda_i \int_{\bar{\gamma}^i}^{\infty} \gamma^i dF(\gamma^i). \tag{2.22}$$

Interpretation: Given the strategies of the other firms, firm i is indifferent between entering directly with a realized innovation of quality $\bar{\gamma}_i$ and undertaking one more research trial.

An equilibrium in threshold strategies will be a solution of the system of equations (2.22) for each firm. It can easily be verified that an increase in the threshold of any firm other than i leads to an increase of the optimal threshold according to (2.22). This together with an algebraic fixed point theorem implies the existence of equilibrium.[16] Moreover, it allows for quite simple comparative statics.

If the hazard rate $f(\gamma)/(1 - F(\gamma))$ is nondecreasing everywhere, the equilibrium is unique.[17] Let me assume that this is the case. If an increase of a parameter leads *ceteris paribus* to an increase in the optimal threshold according to (2.22), the threshold chosen in equilibrium increases as well. Moreover, also the optimal thresholds of all other firms increase. Consequently, the equilibrium thresholds of the firms are increasing in the efficiency parameter λ_i, and decreasing in c_i, in r, and in the number of firms. If firms differ only in their research costs, a firm with a lower research cost chooses a higher threshold level. If firms only differ in their efficiency parameters, more efficient firms choose higher threshold levels.

Example Continued

Let all firms be identical. They choose their threshold rules according to equation (2.22):

$$(r + (n-1)\lambda e^{-\bar{\gamma}})\bar{\gamma} = -c + \lambda e^{-\bar{\gamma}} \tag{2.23}$$

The equilibrium is always unique due to the fact that the hazard rate is everywhere constant in the case where the outcomes of the research trials are exponentially distributed. Furthermore the equilibrium threshold decreases with the number of

[16] See Appendix (A.1) for further details.

[17] See Lippman and Mamer [1993] for a proof of this assertion.

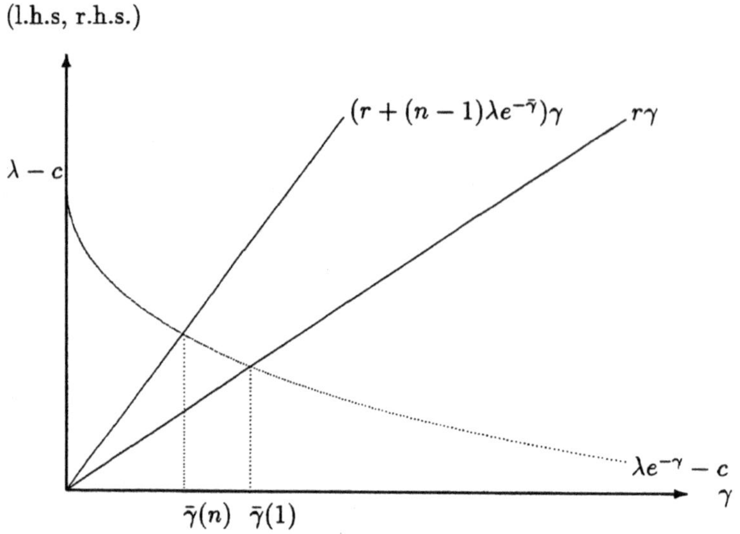

Figure 2.2: The Equilibrium Threshold as a Function of n

active firms in the industry. This is depicted in Figure 2.2. If the discount rate r increases, the left hand side of equation (2.23) is steeper in γ. Therefore, the equilibrium threshold decreases. If the research costs increase, the intersection of the right hand side of (2.23) with the vertical axis moves downwards. Again the equilibrium threshold decreases. An increase in the efficiency parameter λ leads to a movement in the opposite direction, thus the equilibrium threshold increases.

2.2.3 Post-Innovation Competition

There are two basic approaches towards modelling one-shot technological competition together with post-innovation competition in a two-stage model: A deterministic approach, and a stochastic one. Both approaches deal with (cost reducing) process innovations.

In the deterministic approach (e.g. Dasgupta and Stiglitz [1980b]) the setup is as follows: The first stage is similar to an all-pay auction where the bids correspond to the R&D-expenditures each firm is willing to pay. The second stage is a market competition game. The formulation of the first stage differs from a classical all-pay auction in that all firms get a pay-off on the second stage, and there are externalities from market competition. The number of firms is either exogenously given (version 1), or endogenously determined by a free-entry condition (version 2). Dasgupta and Stiglitz analyze Cournot competition in the second stage. With free entry, they derive a linear relationship between the degree of concentration in an industry[18] and the industry ratio of R&D expenditures to sales. The factor of proportionality is given by the inverse of the elasticity of demand. They thus develop a descriptive theory of the variation of R&D-intensities with respect to sales across industries: The intensity should vary positively with the degree of concentration as well as with the elasticity of demand. The more elastic demand, the more concentration leads to an increased R&D/sales-ratio and vice versa.

With exogenous barriers to entry (version 1), increased competition leads to decreased investment by each individual firm, while total industry expenditures are increasing with the number of firms. Thus, competition leads to wasteful duplication of research effort as in the simpler model with perfect patent protection discussed in the last subsection, while at the same time increased rivalry reduces individual incentives, since research expenditures are strategic substitutes (committed *ex ante*).

A somewhat different approach reflecting the uncertainty inherent in innovative activity can be found in Reinganum [1983a]. This model is the first search theoretical model of technological competition in the literature. The game again proceeds in two stages. At the first stage, firms independently evaluate some out of a finite number of research projects $i = 1, \ldots, K$. The outcome of each project is a random variable associated with a constant marginal cost of production c, drawn from an interval $[\underline{c}, \bar{c}]$. The outcome is random, but each firm knows the distribution of potential outcomes for each individual project it may pursue. Firms choose the order in which they evaluate the projects, and after each trial they decide whether to evaluate another one or to cease research activity. The costs for each trial are allowed to vary across projects and across firms. After all firms have decided to cease development, the final production costs are anounced, and the firms enter the second stage. At this stage firms engage in Cournot competition as in the model of Dasgupta and Stiglitz.

An equilibrium for the first stage is thus a *selection rule* about the order in which

[18] The concentration index is given by $\frac{1}{n^*}$, where n^* is the equilibrium number of firms in version 2.

the projects are evaluated, and a *stopping rule* specifying the (random) number of projects evaluated for each firm which cannot be improved upon individually by unilateral deviation.

Without giving a formal description of the model, I state the main results. Equilibrium is completely characterized by a list of thresholds for each firm determining when to stop the research activity.[19] The firm stops doing research after it finds the first technology with a production cost lower than a certain threshold. The threshold of each individual firm is decreasing in the thresholds of the other firms due to the strategic substitutability coming from the assumption that post-innovation competition is Cournot.

2.2.4 Discussion

The simple one-shot patent race has several shortcomings. First, the assumption of perfect appropriability of returns is rather unrealistic in general, even though it might be appropriate for special cases. The empirical literature on technological competition emphazises the importance of technological spillovers.[20] The effectivity of patents is low in many industries, which calls for a theory explaining the choice of means to protect rents from innovations.[21] After all, in most industries other than chemicals and pharmaceuticals there are no effective means for rent protection,[22] and thus there will be subsequent entry of rivals and post-innovation competition.[23] A descriptive theory should therefore take into account how subsequent entry alters the profits of first movers, and how subsequent entry takes place.

[19]In addition every firm chooses a selection rule for the order in which the projects should be evaluated. An optimal order of evaluation is determined by the rule "evaluate the project next that leads to the largest expected profit increase (contingent on the strategies chosen by rivals) relative to its cost." Under the (quite unrealistic) assumption that more costly projects are dominated (in the sense of FSD) by cheaper projects the optimal order due to this rule can be determined *ex ante* (simply begin with the cheapest that leads to the largest expected cost reduction), and therefore plays no further role in the analysis.

[20]For extensive evidence see Levin, Cohen and Mowery [1985], Levin e.a. [1987], Mansfield, Schwartz and Wagner [1981], or Mansfield [1985].

[21]Levin e.a. [1987] stress this point as being of fundamental empirical relevance.

[22]See Mansfield, Schwartz, Wagner [1981].

[23]For the typical time path of the evolution of market structure after a successful innovation, see Gort and Klepper [1982]. They analyzed 46 product innovations in different industries.

The first part of the last argument is the starting point for the two-stage innovation models. There is post-innovation competition in the product market, and the form of competition shapes the incentives for innovation at the first stage of the game. Nevertheless, entry is simultaneous, and rivals cannot react to the technology choice of early innovators. It is endogenously determined, how post-innovation competition takes place, as is the effect on the technology choice of firms in the first place.[24]

In Chapter 3, I develop a framework building on the model of Lippman and Mamer [1993] where late innovators (or imitators) may react to the choices of early movers. This integrates the theory of competition and the theory of innovation incentives in a single framework, and highlights the mutual dependence between the two.

2.3 Dynamic and Multi-Stage Technological Competition

While the first static models of technological competition are concentrated on the question of how many resources firms devote to the achievement and realization of a single innovation, more recent dynamic approaches are used to analyze a larger set of issues. These include the evolution of oligopoly over time, the differential incentives for firms in asymmetric settings, persistence of monopoly and increasing dominance, or the dynamic implications of spillovers for the incentives to innovate and the timing decisions.

Especially the hypothesis that "success breeds success"[25] in technological competition has provoked many theoretical treatments in dynamic settings.

In addition, a vast number of empirical studies analyzed the question whether big firms are more innovative than small firms, or whether concentrated industries tend to produce more innovations. A bulk of Schumpeter-, Neo-Schumpeter, or even Neo-Neo-Schumpeter hypotheses entered the discussion, and still, there is no clear

[24]This is the dark side of the medal, or as Shapiro [1988] puts it: "Two-period models have the advantage of being quite tractable while highlighting the importance of timing. They suffer, however, from their artificial timing structure: The second and final period is essentially one of static oligopoly..." (p. 382).

[25]This phrase was coined by Phillips [1966].

evidence.[26]

2.3.1 A Sequence of Drastic Innovations

Gilbert and Newberry [1982] analyze a situation where a current monopolist is challenged by potential entrants. Entry is only possible through a patented innovation. The current monopolist holds a perfect patent for the product he is currently producing, but by developing a related innovation which is itself patentable, an entrant can actually invent around the patent of the incumbent without inflicting her patent rights.

Assume that the environment of the industry is essentially static and well known to both the incumbent and potential entrants. There is a number of possible innovations an entrant could make to enter the market. At the same time the current incumbent might choose to make the innovation himself, patent it and hold it as a *sleeping patent.*[27] He thus has to decide whether to fight entry by pre-emptive patenting or to accomodate entry .[28]

The game provided by Gilbert and Newberry is static and deterministic. The incumbent holds a patent for the current production technology. There is an alternative technology available. Either the incumbent or a potential entrant may buy the technology and patent it. The setup is an auction where the technology is sold and the question is who will quote the higher bid, the incumbent or the entrant. There is a *joint profit* or *efficiency* effect working, leading to the result that the monopolist will quote the higher bid (with no bid by entrants in an all-pay auction) whenever monopoly profits are higher than joint duopoly profits in the case where the challenger gets the innovation. The reason is that the monopolist may bid just above the expected future profit of the entrant (and thus wins the auction for sure), and still receives a higher net payoff from preserving her monopoly position than she would receive from accomodating entry. Thus there is persistence of monopoly exactly in the case where it is efficient.

[26]For a review of the empirical literature see Cohen and Levin [1989].

[27]The firm does not use the patents that it only holds to prevent rivals from *patenting around*, since it has no incentive to replace itself.

[28]This reminds of the classical chain-store game. In a repeated setting with some uncertainty about the incumbent's type, reputation effects might enter and lead to deviations from the joint-profit maximizing results derived below. See Milgrom and Roberts [1982].

In a now classical paper, Reinganum [1985] reversed the result of Gilbert and Newberry using a dynamic model of industry evolution with drastic innovations. She highlights an effect neglected by Gilbert and Newberry. There is a number of $T+1$ subsequent drastic innovations generating monopoly profit flows $V_0, V_1, \ldots, V_T$, denoted by index t. At each stage, a given number n_t of firms is competing for the innovation. The firms decide on the R&D-expenditure rates x_{it}, i.e. there is no contractual commitment.[29] The constant discount rate is r. Effectively, the firms play an $(T+1)$-repeated Lee and Wilde patent race. The same technological assumptions are made as in Section 2.2.1. The research technology exhibits initially increasing returns to investment up to a threshold spending-level, and decreasing returns thereafter.

The game is solved by backward induction: The last stage is the traditional one-shot Lee and Wilde game. The value for the winner (subscripts supressed), $v^w = V/r$, is greater than the value for a loser $v^l = 0$. From this, one derives by induction that on every stage the continuation value for the current incumbent is higher than for the current challengers ($v^w > v^l$). On every stage firms maximize their expected profits by choosing their expenditure rate. Let w.l.o.g. firm 1 be the current incumbent. The firms solve:

$$\max_{x_1} V^1(x_1, a_1) = \frac{h(x_1)v^w + a_1 v^l + V - x_1}{r + h(x_1) + a_1} - F, \qquad \text{and} \tag{2.24}$$

$$\max_{x_i} V^i(x_i, a_i) = \frac{h(x_i)v^w + a_i v^l + x_i}{r + h(x_i) + a_i} - F. \tag{2.25}$$

From this the first order conditions are derived:

$$h'(x_1)\frac{x_1 - V + (a_1 + r)v^w - a_1 v^l}{r + h(x_1) + a_1} = 1, \qquad \text{and}$$

$$h'(x_i)\frac{x_i + (a_i + r)v^w - a_i v^l}{r + h(x_i) + a_i} = 1.$$

It is assumed that the second order conditions are strictly fulfilled, i.e.

$$h''(x_1)(x_1 - V + (a_1 + r)v^w - a_1 v^l) < 0, \qquad \text{and}$$

$$h''(x_i)(x_i + (a_i + r)v^w - a_i v^l) < 0.$$

If the sufficient conditions are strictly negative it can be concluded from the first order conditions that $h''(x_i)$ is negative for all challengers. Thus, all challengers overinvest (compared to the benchmark of a monopolist or social planner. See the basic

[29] Thus, R&D-expenditures are strategic complements by the argument developed in the last section.

static model in Section 2.2.1) at each stage. The same is not necessarily true for the current incumbent. It is clear from the first order condition that the current incumbent invests less than any challenger does. This is the *replacement effect* neglected in the model of Gilbert and Newberry. This effect is dominant in the present model for two reasons: First, the *joint profit effect* is zero, because innovations are drastic and research expenditures are non-recoverable. Second, despite the fact that the current incumbent loses his current monopoly profits if he does not win the current stage, he has less incentives to invest. This is so because of the differential gains accruing to the incumbent and the challenger from winning the current stage, and because of the uncertainty concerning the date of successful completion of the next innovation, as was pointed out by Reinganum [1983b, 1984]. Note that uncertainty contributes to the replacement effect, but is not a necessary constituent.[30]

Vickers [1986] analyzes a similar framework in which there is also a sequence of innovations. At each stage there is a deterministic second-bid all-pay auction for an innovation. There are two firms. In the case of drastic innovations there is increasing dominance and persistence of monopoly. If production costs are reduced by each innovation, each firm's profit depends on the number of innovations it has completed. Under the assumption that joint profits are maximized if the firms do not differ very much in the number of innovations, there is a joint profit effect working, which leads to an *action-reaction* profile where the current loser will win the next auction.

An interesting by-product of the analysis of Vickers is the following. If the firms market their innovations directly and are engaged in Bertrand competition this reduces the speed of research undertaken by the firms, while Cournot competition increases the speed of research. A parallel result can be found in Mookherjee and Ray [1991]. In their model the technological knowledge of the innovator gradually leaks out to a competitive fringe of the market.

An important effect which drives the different results is the uncertainty of innovations and their drasticity introduced in Reinganum's approach. Thus, if there is a sequence of drastic innovations, and some uncertainty about the evolution of the industry in the future, one should not observe increasing dominance. I will discuss another study by Budd, Harris and Vickers [1993] in the next section, where all

[30]See the discussion in the introduction of this chapter, and Lippman and Mamer [1992].

effects can be analyzed in an integrated framework with incremental innovations.

The assumption of a sequence of drastic innovations with post-innovation monopoly is quite artificial. Radical or drastic innovations do not occur that often, and in the time between drastic innovations there is imitation and competition between a number of firms in a given industry. Mookherjee and Ray [1991] analyze a model where there is a series of innovations available to a big established firm, but technological knowledge gradually leaks out to firms at a competitive fringe of the market. These spillover effects influence the time spacing between the introduction of subsequent innovations. The pace at which sequential introduction takes place, and the effect of the rate of diffusion of knowledge to rivals depend crucially on whether firms compete in prices or quantities in the product market. Under price competition the innovations are spaced away optimally, while under quantity competition all innovations that are adopted at all, are adopted instantaneously.[31] In addition, there is a tendency to stop the adoption of innovations before all possible innovations are introduced which may be strenghtened by increased diffusion both under price and quantity competition.

2.3.2 Evolution of Oligopoly with Incremental Innovations

Giving up the assumption that all innovations are drastic leads to the analysis of the dynamics of an established oligopoly if there are incremental improvements of technology or gradual differentiation in the product market. This approach is perhaps best termed *analyses of optimal adaptation*, since not the single identifiable innovation but the fine-tuning of production technology or product positionning in a dynamic setting is of interest. I review two studies which analyze this phenomenon, Võlker [1990], and Budd, Harris and Vickers [1993].

Consider first the search theoretical approach of Võlker [1990]. There are two firms competing in a given product market. Time is discrete. Each period the firms decide

[31]This effect comes from the sign of the cross derivative of the established firm's profit function with respect to own and rival's marginal cost of production (whether cost levels are *strategic complements* or *substitutes* for the established firm). The sign shapes the change in the incentive to innovate while rival's cost decrease. The authors assume an increasing (decreasing) incentive to innovate while rival's cost decrease in the case of price (quantity) competition. Thus the diffusion of (cost reducing) innovation does not create a higher incentive to innovate for the leader in the case of quantity competition. Therefore, all innovations that are adopted, are adopted instantaneously.

whether or not to undertake a research trial. Each research trial has a fixed cost F, and leads to random realization of a technology level y from the unit interval $[0,1]$. Realized improvements can be directly implemented into the production process without any further costs. The c.d.f. for the technology realizations on the unit interval is denoted by $F(y)$ which is continuous and has a p.d.f. $f(y)$. Per-period profit functions $\Pi_i(y_i, y_j)$ are twice continuously differentiable. Profits are increasing in the own technological level, while decreasing in the technological level of the rival. The analysis is undertaken separately for the cases of *strategic complementarities* and *strategic substitutabilities* in the technology levels used by the firms.

Firms decide independently from each other whether to invest into R&D or not. Suppose the current state of technology is (y_i, y_j). The optimal strategy of each firm can be characterized as follows: For each technology level y_j of the opponent there exists a threshold level $\bar{y}_i(y_j)$. If the current technology of firm i is more advanced than this threshold level ($y_i > \bar{y}_i(y_j)$), the firm stops investing into R&D. Otherwise it will undertake another trial. Furthermore, the thresholds are increasing in the technology level of the opponent in the case of strategic complements (by definition), and decreasing in the case of strategic substitutes. The slope of the threshold curve in the (y_i, y_j)-space depends on whether the rival firm still invests in R&D or not.

The following dynamic behavior emerges from the model: If the technology level of firms is low, both firms invest. A firm which is far more advanced than its rival stops investing. The exact stopping region depends on the strategic situation in the product market. A firm stops "earlier" if profit functions exhibit strategic substitutabilities. If the gap is narrowed by the laggard, the leading firm may re-enter investment, as long as the profit functions are not strategically independent.[32] Both firms stop if they are relatively advanced. In most cases this will happen before the technological limit is reached.

The firms choose to stop investment in some cases because of the discrete nature of the game. If investment levels were variable the intensity of investment would change with the state competition is in. To demonstrate this effect, I turn to the continuous-time framework of Budd, Harris and Vickers [1993].

Again, the situation of duopoly is analyzed. Instantaneous profits depend on the *technology gap* between the two firms. Let the technological levels of the two firms

[32]This is the "reswitching property" first analyzed by Lee [1984] in stopping games.

be denoted z_i and z_j. The technology gap $s = z_i - z_j$ lies in the unit interval $[0, 1]$ by definition. The firms earn instantaneous profits $\pi(s)$ and $\pi(1-s)$. At every instant in time each firm decides on its current R&D-efforts, $x_i, x_j : (0,1) \to [0, Z]$, where Z defines some upper bound on efforts. The associated effort cost functions are $c(x_i)$, and $c(x_j)$ respectively.

The law of motion is represented by a standard Wiener process. The drift is given by $(x_i(s) - x_j(s))dt$, and the variance by $\sigma^2 dt$. At the boundaries the law of motion is parameterized with a *repulsion factor* λ, i.e.

$$ds = \frac{1}{2}\sigma^2 \lambda dt, \quad \text{at } s = 0 \quad ds = -\frac{1}{2}\sigma^2 \lambda dt, \quad \text{at } s = 1. \tag{2.26}$$

A repulsion factor of zero leads to an absorbing state at the boundaries. This is the usual patent race. Once the boundary is reached, monopoly is installed and there is some flow rent from the patent. With a finite, but positive repulsion factor the process rests at the endpoints with some positive probability. This may be interpreted as a transitory monopoly where the technological advantage stochastically leaks out to a potential competitor. The discount rate r is assumed to be positive.

The model is solved by dynamic programming methods. Suppose play is in some state s. The expected future payoff is denoted V. We have (for firm i)

$$V(s) = r[\pi(s) - c(x_i(s))]dt + (1 - rdt)EV(s + ds) \tag{2.27}$$

Expanding EV around s, ignoring higher order terms leads to

$$EV(s + ds) = V(s) + V'(s)(x_i(s) - x_j(s))dt + \frac{1}{2}\sigma^2 V''(s)dt. \tag{2.28}$$

The last two equations together imply

$$\frac{1}{2}\sigma^2 V'' + (x_i - x_j)V' + r(\pi - c(x_i)) - rV = 0. \tag{2.29}$$

The optimal control $x_i(s)$ thus maximizes $[x_i(s)V'(s) - rc(x_i(s))]$. Thus,

$$c'(x_i(s)) = \frac{V'(s)}{r}. \tag{2.30}$$

It is assumed that the optimal choice $x_i(s)$ is unique. Let the value for the second firm be denoted $W(s)$. By assuming $c'(.) > 0$ one has that $[x_i(s) - x_j(s)]$ has the same sign as $[V' + W']$. From this it can be concluded that competition moves into the direction where joint profits increase.

Some other effects are verifiable in the current model if the discount rate is high. The first Budd, Harris and Vickers call the *profit-incentive cost effect.* It says that a firm works harder if its profit function is increasing and concave in the state, or if rivals profits are decreasing and convex. The second effect is the *endpoint effect* which depends on λ. If there is absorption at the barriers, spending stops once a barrier is reached. If there is infinite repulsion, effort of both firms slows down while the state approaches the boundaries. For intermediate repulsion both effects are working.

Many interesting facettes of dynamic technological competition are captured in the two models of this section. Even though they are restricted to the stylized framework of duopoly they provide elegant analyses concerning increasing dominance and differential incentives in response to changing market conditions. An obvious drawback is the costless and instantaneous implementation of innovations. Additionally, in Völker's model many results are consequences of the discrete structure of the model (which nevertheless seems to be reasonable in many cases). In Budd, Harris and Vickers the process of innovation is rather abstract. There is a flow of incremental innovations, no delay between spending and realization of the results, and all the uncertainty is exogenous.[33] Especially the last point deserves discussion, since it is not at all clear whether the uncertainty stems from the research technology, from competitors, or from general market conditions. In so far, the approaches discussed earlier are more sophisticated and advanced (even if static in nature), and correspond more to what one has in mind about how the real world process of innovation.[34]

[33]There are two more questionable points, namely the non-standard distributional assumption on the innovation process (the joint process is normal), and the unidimensionality of the state space. The last point is no serious restriction in the game with positive repulsion at the boundaries, since in this case the game never ends. Having two dimensions for the two firms means that, beside a relative position compared to the rival, in addition each firm is characterized by a relative position compared to the end point. This contains no additional information in the infinite game.

[34]The models of Loury, Lee and Wilde as well as Dasgupta and Stiglitz allowed for a separated analysis of the implications of uncertainty coming from rivals (market uncertainty represented by the random minimum completion time of the other firms) and own technological uncertainty. See Section 2.2.1.

2.3.3 Multi-Stage Research

In most of the approaches presented in the previous sections there is complete information. Once an innovation is made, it is implemented and observed by the rival. In a multi-stage research framework, the informational structure determines the incentives to complete the task. If a firm proceeds faster than its rival, the rival may want to give up research before completion, because she sees no chance to catch up. This consequence is largely dependent on two facts. First, uncertainty in the process preserves the chance of catching up. If the game has the form of a race[35], uncertainty will make laggards race longer. Second, the informational structure determines what the reaction of firms on each other can be conditioned upon. This in turn will influence the decision whether to continue investing or to abandon research.

Special features of multi-stage races are that they allow for the analysis of dropout without success, the pace of progress under different informational assumptions, and the possibility of overtaking or *leap-frogging*. I will give a short digression on four approaches from the literature in the sequel, namely Fudenberg *e.a.* [1983], Harris and Vickers [1987], Grossman and Shapiro [1987], and Lippman and Mc Cardle [1987].

The models fit into the following general framework. There are two firms $i = 1,2$ competing for an innovation. If firm i wins the race it gets a price V_i^+, and it receives V_i^- if it loses. All models provide examples of a two-dimensional state-space $\mathcal{S} = S_1 \times S_2$. There is a total of N stages to be solved before the price is awarded. If firm 1 solved j stages, while firm 2 solved k stages, the game is said to be in stage (j,k). Call the subgame in that stage $G(j,k)$. In Figure 2.3.3 I provide an example for such an state-space. On every stage the firms may decide whether to stay in or to drop out of the race. In the models by Harris and Vickers [1987], as well as Grossman and Shapiro [1987] they are also allowed to choose their stage R&D-intensity. I turn to the analysis of the different approaches.

The first approach is asymmetric in nature. In the model of Fudenberg *e.a.* there are two firms with the same deterministic constant returns-to-scale learning technology. Time is continuous. The firms accumulate knowledge over time while spending on R&D at a constant rate. The higher the knowledge level of a firm the higher its

[35]This is the case whenever the expected profits of the first innovator are higher than the expected profits of followers. See the model in Chapter 3 for details.

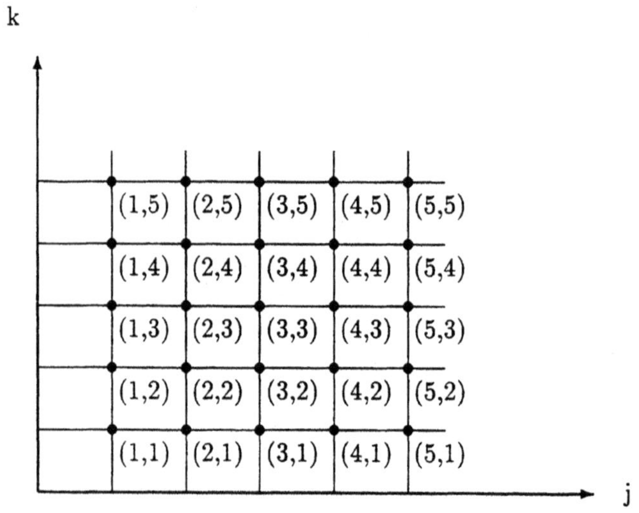

Figure 2.3: The State-Grid of a 5-Stage Patent-Race

instantaneous success probability. The firm to complete the innovation first gets the whole surplus. A monopoly will find it profitable to continue research while it is not profitable for both firms to always undertake research. One firm is assumed to have a headstart over the other firm. In this case it never pays for the other firm to start investing. This phenomenon is called *ϵ-preemption*.

To understand *ϵ-preemption*, look at the symmetric case which is the classical *war of attrition*. In equilibrium there is perfect rent dissipation and thus both firms make zero profits in expected terms.[36] If one firm has a whatever small headstart the laggard cannot catch up in the constant investment model. Thus, expected profits for the laggard are always negative if she starts to invest into R&D.

In a multi-stage version with lagged monitoring, a firm drops out if it is more stages behind than the length of the revelation lag. If the lag is smaller, the laggard mixes between staying and dropping out. If both firms are tied and the endpoint is in reach, they stay in with certainty.

Lippman and Mc Cardle allow the firms to differ in their (fixed) *spending rate*

[36]See Fudenberg and Tirole [1991] for an explicit derivation of the rent-dissipation result.

c_i, as well as in their *efficiency* parameterized by different hazard rates λ_i (stage completion arrives exponentially). Play proceeds through N stages. Final payoff is V_i under a winner takes all rule. The discount rate is set to zero.

The expected completion time has thus a gamma distribution with parameters (N, λ). It is assumed that a single firm never stops, and that it is not profitable for both firms to commit to staying in the race until the innovation is completed. The game is played under complete information (perfect monitoring).

The following results can be obtained. If firms are symmetric, the leader never drops out. The laggard drops out if the lag is large. The dropout region is decreasing in the own valuation of the game V_i, and in efficiency λ_i, while increasing in own research cost c_i. Furthermore, it is increasing in the rivals' valuation of the game V_j as well as in its efficiency λ_j, and decreasing in rivals' research cost c_j. The results are obtained by exploiting the algebraic structure of the strategy space.

A more elaborate model can be found in Harris and Vickers [1987]. The stage game is like in Lee and Wilde [1980], but reduced to the case of duopoly. The fixed costs are normalized to zero, and current R&D-spendings are convex in innovation effort.[37]

The main purpose of the paper is to analyze the incentive differentials between firms as competition evolves. Like in the continuous model of Budd, Harris and Vickers [1993], the authors characterize the conditions under which firm 1 exerts higher effort than firm 2. Again, there is a *joint-profit effect* working: Firm 1 exerts higher effort than firm 2 if and only if the joint continuation profit is rising conditional on firm 1 winning the game in the current stage. The driving force behind this effect is the size of the relative (potential) losses the firms may experience by losing the current stage game.

A simpler two-stage version of Harris and Vickers [1987] with positive discounting can be found in Grossman and Shapiro [1987]. Again the stage game is taken from Lee and Wilde. The firms invest more if they are tied. A tie on the second stage leads to more vigorous competition than a tie on the first stage. The follower will slow down or even cease investment if the leader steps ahead. If, by chance, the

[37] Remember that in the basic Lee and Wilde version expenditures are linear in effort, while the hazard rate is s-shaped. The Harris and Vickers assumption of convex cost corresponds to a hazard rate that is concave in effort (i.e. $h''(x_i) < 0$).

follower catches up, the leader speads up research. Positive discounting can lead to spurred investment if the follower drops out, while this is never the case with zero-discounting.

2.3.4 Discussion

In the last subsections I presented a variety of different approaches used to analyze the dynamics of technological competition. There are two broad themes. First, is there *increasing dominance* of a firm, and second, how does *competition shape technological progress* within an industry?

Consider first the *increasing dominance* debate: There are two major effects working. First, the *replacement effect* introduced by Arrow [1962], and highlighted by Reinganum [1983, 1985]. The effect is dominant if the innovations are relatively drastic, and if there is sufficient uncertainty in the model. The second effect is the *efficiency* or *joint-profit* effect. This effect is prominent in all auction models where only the highest bid is forfeited and in models with no uncertainty. It does not appear in Reinganum's model, since there is a series of drastic innovations. The first effect stresses the relative gains for the firms from moving ahead, while the second effect sheds light on the relative losses the firms may experience. As pointed out by Lippman and Mamer [1992], a common source of the two effects lies in the relative size of winner's and loser's profits from technological competition.[38]

A third effect which is prevalent in most of the models, but not always identifiable, comes directly from strategic considerations. Many results depend on whether the strategies are substitutes or complements. In the one-shot models and in the discrete state-space models with only two strategies in each state ("continue", "stop") those effects can be traced easily. This can clearly be seen in the models of Völker or Lippman and Mc Cardle. The search model of Reinganum has this feature, too. The strategic structure of the game does not necessarily determine the differentials in incentives coming from asymmetries in the setting, but they shape the structure of the aggregate incentives of the firms to invest in a specific situation.

The more elaborate approaches are shaped by the same tradeoffs, but these tradeoffs turn out to be much more difficult to trace. I tried to find a relation between

[38]See also Chapter 3 for a further analysis of the implications of the relative size of profits.

the apparent effects in the Harris and Vickers-model and the notion of strategic complements or substitutes, but the characterization is impossible without further specification of the game. Harris and Vickers, as well as Budd, Harris and Vickers used simulation methods to gain further insights into the mechanics of their model. But this bears also a serious danger: Budd, Harris and Vickers derived a number of effects in their model by taking different approximations as well as by simulation studies, but they are no longer able to trace back the effects to their sources. Moreover, it is not at all clear whether there are different effects working in the model, or if the authors just identify different aspects of one underlying effect.

The major step in modelling the dynamics of R&D investment was the introduction of the state-space approach. It allows for the most flexible "who knows what" specification. Despite this general flexibility of the approach, it is rarely used. Most of the models have either incomplete information during the whole research stage (which is then effectively a single stage), or they specify perfect information, i.e. the new state is always revealed to all players after each player's move. In Chapter 4 I provide an analysis of the incentives firms might have to strategically influence the information structure of the game. This question has recently attracted increased interest in both the theory of R&D and technological competition as well as in the theory of investment.

2.4 Sequential Entry and Technology Choice from a Given Set

How does the division of profits in the post-innovation market shape the innovative activity of firms, and how in turn does the division of profits arise from that activity?

There are not yet many authors, who tried to develop a theory in this direction. After all, there are three approaches incorporating technology choice and sequential entry. The first two, Dasgupta [1986, 1988], and McLean and Riordan [1989], restrict the analysis to the decision whether to innovate or to imitate, or to the choice from a set of given technologies. The third model is Dutta and Rustichini [1993]. This model is more related to my model, and therefore I confront their findings to mine in the next chapter.

A\B	enter now	enter later
enter now	$(4, 4)$	$(5, 10)$
enter later	$(10, 5)$	$(8, 8)$

Figure 2.4: An Innovation/Imitation Game with Second-Mover Advantages

Mansfield, Schwartz and Wagner [1981] show in an empirical investigation that technological knowledge leaks out from first-movers to later imitators. Therefore, imitation costs are often lower than innovation costs. In the empirical study an average ratio of 0.65 was calculated. Dasgupta [1986] takes this ratio and concludes that there may well be an advantage for imitators. Suppose that imitation costs are much lower than innovation costs, and in addition, knowledge leaks out quite rapidly. There are no effective barriers to entry in the respective industry. The strategic structure of such an situation can be represented as a game as depicted in Figure 2.4.

The game in 2.4 has two asymmetric equilibria in which one firm innovates and earns low profits, and the other imitates and earns high profits.

Dasgupta [1988] coined the term *waiting game* for this case to distinguish it from the patent *races* discussed earlier. In a race every firm wants to be the first in the market (i.e. in the case of perfect patent protection), while in a waiting game every firm wants to be the follower.

McLean and Riordan [1989] take another perspective: There are two technologies, α and β. The α-technology guarantees an overall higher profit, wheras the β-technology is better qualified to deter further entry. One might think about Cournot-Nash competition with constant marginal costs, where the marginal cost of the α-technology exceeds the marginal costs of the β-technology, and where the inverse relationship holds for fixed costs. If the fixed cost of the β-technology sufficiently high, we can observe two effects: A firm choosing the β-technology deters further entry more effectively, but makes lower overall profits in the relevant range. Entry is "sequential" in that the firms move sequentially on the technology choice stage. Thus, the first firm chooses its preferred technology, then the second, the third firm, and so on. On the last stage there is a simultaneous-move market competition game.

Under these conditions there exists an equilibrium where the first set of firms chooses the α-technology, the second set of firms chooses the β-technology, and the remaining firms stay out. Thus, there is always a first mover advantage (as expected from the structure of the game).

To formulate the tasks still unsolved, I give a quote from McLean and Riordan (p. 20f):

> "A desideratum for a complete theory of industry organization is an explanation for the evolution of industry structure, including the determination of entry and cost structures [...]
>
> An important assumption of our analysis was that firms' equilibrium profits depended only on technology choices, and not on the firm's position in the entry queue. We interpret this to mean that entry lags are infinitesimal small. However, with an expanding market or technological change, entry might be spread over time."

This defines quite accurately the program for the next chapter. First, I analyze the endogenuous timing decision and the technology choice upon first entry and second, I concentrate on the spacing of subsequent entry and its consequences for innovation incentives.[39]

[39]The case of stochastically increasing demand has been treated in a paper by Glazer [1985]. The shortcoming of this paper is the concentration on the timing decision, while the distribution of profits between first movers and second movers is assumed to be exogenous.

Chapter 3

Sequential Innovative Entry and Endogenous Technology Choice

3.1 Introduction

In this chapter I develop a dynamic model of *technological competition* in which the choice of technology is endogenized. The natural starting point for such a model is the framework of Lippman and Mamer [1993]. In addition to the endogenization of technological standards chosen by firms under competition, I allow for sequential entry into the market. The entry times are freely chosen by the firms. With this generalization I obtain different classes of games of technological competition. The Lippman and Mamer model turns out to be a special case in the class with *first-mover advantages.* There may also be endogenously arising *second-mover advantages.*

Motivation

My interest is in the interaction between innovative activity and industrial structure. What determines the incentives to innovate? – Let me summarize the two positions formulated in the theoretical literature:

- *Present market structure:* This strand of the literature stresses the difference in incentives between current incumbents and current challengers. The current challenger has more to win, while the current incumbent has more to defend.

These are the sources that shape the incentives to invest. The literature into this direction was reviewed in Chapter 2.3.

- *Expected market structure:* This strand of the literature can grossly be subsumed under the term *patent races.* The question is, how much a firm is willing to invest to achieve a dominant position in the post-innovation market. See Chapter 2.2 for details. Fewer papers are devoted to the question how the order of entry and the division of profits in the post-innovation market between several innovators shapes the incentives to invest (See Chapter 2.4).

In the present chapter I follow the second strand of reasoning. I do not claim that the first topic is irrelevant, but I have the impression that it is quite well understood. Therefore it is the second topic that I am more interested in. In the literature so far, the choice of quality or technological standards is either exogenous, or it is assumed that the post-innovation market is monopolistic. In most of the models that treat competition in the post-innovation market at all, entry is simultaneous and therefore the choice of technology and the division of profits is not influenced by the order of entry into the market (or by the entry time). The others have a predetermined entry-order, or simplify the analysis to an innovator/imitator-game.[1]

Hypotheses and Questions Addressed

Let me start with some hypotheses:

- The order of entry is an important determinant for the expected profits of firms.
- There are time lags between the market entry of different firms.
- The fact that there may be subsequent entry into a market influences the innovative activity of firms.
- The technological standards (or product qualities) already used in the market shape the innovative activity of later entrants.
- From the last two points it should be clear that the division of profits in the post-innovation market is jointly determined by the strategic choices of early and late entrants into the market.

[1] There is one exception. Dutta and Rustichini [1993] present a model that is closely related to the one developed here. It will be discussed later.

- Research is conducted to differentiate from rivals. Thus, there should be an important idiosyncratic component.

- Innovative activity differs from pure investment in that it is risky. The outcome as well as the duration of a research project is therefore likely to be random.

Discussion: The first point is prominent in many empirical studies of innovative behavior. Mansfield *e.a.* [1981] and Mansfield [1985] show that technological knowledge of firms active in the market leaks out to rivals. Levin *e.a.* [1987] emphasize the importance of the order of entry. leadtime is often considered to be important for the appropriation of rents from innovation. This is so because legal protection of rents is often considered to be ineffective, and subsequent entry therefore frequently occurs.

Beside the fact that lead time is important, to enter the market with a new product of low quality (*i.e.* because of time constraints) may spoil the market for the early entrant. She may lose reputation or offer a cheap chance for rivals to rapidly enter the market with improved products. To give an example, consider the ulcer-drug industry (see Berndt *e.a.* [1994]). In 1977 SmithKline entered the american market with a new type of anti-ulcer medical product (Tagamet). This opened the way for new therapies for this desease. Nevertheless, because of the early entry of SmithKline only few indications (FDA-approved treatments) were granted to SmithKline by the public authorities. Another drawback was the inconvenient frequency of dosage. In 1983 Glaxo entered with a similar product, called Zantac. This product avoided many of the inconveniencies of Tagamet, and had more indications approved. Today there are four brands in the market, and Zantac has by far the greatest market share.

This example gives also evidence to the third hypothesis. The later entrant observed the advantages and disadvantages of Tagamet and had the chance to enter the market with an improved product. This is a clear case of second-mover advantage.

Another example is the behavior of IBM when Apple entered the market with the first personal micro-computer in 1977. IBM was then the dominant producer of mainframe computers. They announced that the micro-computer market won't have a future. A few years later (1980) they entered themselves with the 8088-personal computer. Although they are only one out of many producers of personal computers today, they were able to set the dominant standard in disk operating systems (in alliance with Microsoft) and gained a respectable market share. The early entrants into the personal computer market (Apple, Commodore and Atari) all had operating systems that are held to be superior to MS-DOS. Nevertheless, they (with the exception of Apple) always had problems to attract enough firms to produce powerful software for serious scientific and business applications. The decision of IBM to wait may well have contributed to the reputation that this firm does not react to every strawfever, and is therefore a serious and reliable partner for both customers and software producers. One may well doubt that the success of IBM would have been comparable if it had entered earlier.

The second to last point is theoretically founded. Bhattacharya and Mookherjee [1986] analyze the question whether firms (compared to a social planner) prefer to choose positively correlated projects. They conclude that "if [...] opportunity sets are symmetric, both society and private firms prefer less correlation *per se*" (p.604). This hypothesis is used to substantiate the point of view that research of firms should be modelled as independent processes.[2]

The final point is just definition. I would argue that the main difference between innovative activity and investment in a theoretical treatment is the randomness that is higher in the case of innovative activity.

With the following model I try to integrate the different hypotheses formulated above. Therefore I provide a *stochastic* framework with *endogenous quality choice* and *sequential entry*. The stochastics serve to analyze the research of the two firms as independent processes. This allows for a more detailed discussion of strategic implications than in a simple *game of timing* (like Dutta and Rustichini [1993]). Especially the close connection between the notion of *strategic complements* or *strategic substitutes* and the presence of *first-* and second-mover advantages is only observable if the research processes of the firms are imperfectly correlated. The endogeneity of the innovation quality together with the possibility of sequential entry provides the link between the market and the research stage. It allows for the analysis of the feedbacks from expected market structure on the incentives to innovate (which in turn shape the actual equilibrium market structure). The remainder of the chapter is organized as follows:

---

[2]There may be spillovers at the research stage. These spillovers can be built into the model as cost-reducing. The respective comparative statics result then applies. The main point is that research in my model is not perfectly correlated as it is for example in Dutta and Rustichini [1993].

Section 3.2 the model is presented and the equibrium analysis is conducted. It turns out that the division of profits between early and late entrants is the main force that shapes the incentives to innovate. The comparative statics are conducted in Section 3.3. In Section 3.4 I discuss the endogeneity of the strategic structure of the model. I summarize and draw conclusions in Section 3.5.

3.2 Endogenous Technology Choice and Sequential Entry

In Chapter 2.2.2 I introduced the Lippman and Mamer [1993] model of endogenous choice of quality in a patent race. The same basic idea of a *stopping game*[3] will be used here. The firms invest into R&D projects of uncertain duration and outcome. The duration of a project of firm i is modelled as a hazard rate process with intensity λ_i. This intensity reflects the firms' idiosyncratic skills to solve a problem. Therefore I call λ_i the *efficiency* parameter of firm i. The outcome of each research project is a random variable γ with an absolute continuous distribution function $F(\gamma)$ and density $f(\gamma)$ over the positive real numbers. Firms conduct a series of subsequent projects until they realized an acceptable γ to enter the market.

Setup: I restrict the analysis to the case of two firms. Call these firm 1 and firm 2, respectively. The firms may differ in their efficiencies λ_i, $i = 1, 2$, and in their research costs c_i, $i = 1, 2$. There is no perfect patent protection. This is the major innovation of the present chapter. In Lippman and Mamer the first firm that entered the market is awarded perfect patent protection and the world ends. I allow for subsequent entry of rival firms.

Information: Firms have private information about the past realizations of own research projects. In addition, they know who is active in the market, and which technology (or product quality) is used by active firms.

Strategies: Firms use stopping times as behavioral strategies at each information set they reach. Given the nature of the research process it seems reasonable to restrict the analysis to that of firms choosing stationary strategies. Furthermore, I restrict myself to the analysis of stationary strategies that can be represented by

[3] A standard reference for the theory of stopping times is Chow *e.a.* [1971].

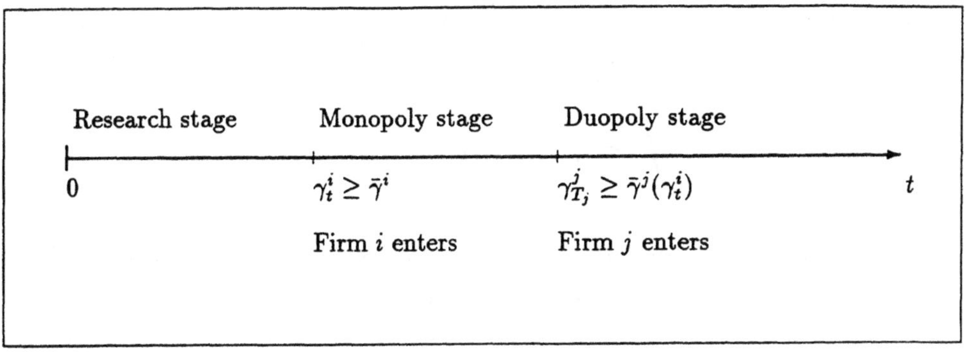

Figure 3.1: Timing of the Sequential Entry Game

simple *threshold rules.* It can be shown that there exists a best response that is representable by a threshold value to every stationary strategy chosen by a rival.[4] Since I allow for sequential entry, an optimal threshold rule is dependent on the state of competition, i.e. the number of active firms and their marketed technology. To put it into formal terms, for firm i, a strategy can be represented by a function $g^i : (\emptyset, \gamma^j) \longrightarrow \mathbb{R}$, where $\emptyset$ denotes the research stage, where no firm is active, and γ^j the realized technology of firm j if it is already in the market. I write $g^i(\emptyset) := \bar{\gamma}^i$, and $g^i(\gamma^j) := \bar{\gamma}^i(\gamma^j)$.

Notation: A *bar* over the value denotes a *threshold value*, while the *value without bar* denotes a *realization* of the technological variable. *Superscripts* denote *strategy variables* of firm $i = 1, 2$, while *subscripts* denote idiosyncratic *parameters* of firm $i = 1, 2$. variables

The game evolves as depicted in Figure 3.1.

Profits: Starting research is profitable for each firm, while it is not profitable to continue forever. I generalize the model of Lippman and Mamer to the case where profits are not identified with the realized innovation, but are related to each other.[5] If a firm decides to stop research in τ after it has realized an innovation

[4]The reason is that the research process is memoryless and stationary. For a proof see Lippman and Mamer [1993], p. 110f.

[5]Identifying the quality of an innovation with its profitability is only possible in a model where the *winner takes all.* Since I concentrate on oligopoly profits are dependent on the entry strategies of all firms and on the emerging market structure.

of technological quality γ^i_τ, it earns an expected profit conditional on the strategies chosen by the other firm which is given by

$$V^i(\gamma^i_\tau, \bar{\gamma}^j) := E\left[\int_\tau^\infty e^{-rt}\pi_t(\gamma^i_\tau, \bar{\gamma}^j)dt\right]. \tag{3.1}$$

Expectations in (3.1) are formed over the stopping time and technological quality of the rival. With $\pi_t(\gamma^i_\tau, \bar{\gamma}^j)$ I denote the instantaneous profit function of firm i. This is the monopoly profit firm i earns on the monopoly stage if it entered before firm j. In this case, $\pi_t(\gamma^i_\tau, \bar{\gamma}^j) = \pi_t(\gamma^i_\tau)$. When both firms are active in the market, they engage in some sort of market competition and earn duopoly profits. Then $\pi_t(\gamma^i_\tau, \bar{\gamma}^j)$ denotes the expected duopoly profits of firm i given that firm j has chosen $\bar{\gamma}^j$ as its threshold strategy.

To guarantee that both firms start research, I impose the following assumption:

Assumption 3.1 *The expected profit from starting research for firm i, $i = 1, 2$ is greater than the flow cost (irrespectively of the strategy employed by the rival), i.e.*

$$\lambda_i EV^i(\gamma^i, \bar{\gamma}^j) > c_i \quad , \forall\ \bar{\gamma}^j \geq 0,$$

where expectations are taken over the realization γ^i of the first trial.

To guarantee a tractable analysis, I impose the next assumption.

Assumption 3.2 *(i) Firms either invest in R&D (and are not in the market) or are active in the market.*

(ii) Once a firm decided to stop R&D and entered the market with an innovation it is from then on committed to this particular product or technology associated with the innovation.

The first part of Assumption 3.2 simplifies the analysis as it reduces the strategy space to a single stopping time. In the next chapter I will investigate the implications of relaxing this assumption in a somewhat simpler model. The second part of Assumption 3.2 may be justified by specific capital investments that render further research unprofitable for the first mover. The argument is related to the replacement effect introduced earlier.[6]

Note that at least the second part of Assumption 3.2 is by no means necessary for the existence or characterization of equilibria. It is made to restrict the analysis to a certain set of interesting questions. Suppose that early movers can adapt to changes in their competitive environment (e.g. to entry of rivals). This would change the relative distribution of first-mover and second-mover profits, and therefore influence the incentives of firms to invest in R&D. These changes in the distribution of profits will be discussed in Section 3.3. Relaxing the first part of Assumption 3.2 leads to the question under which circumstances firms may have incentives to *shelve* innovations. This is adressed in Chapter 4.

I will look for subgame perfect equilibria of the game.[7] To solve for an equilibrium, let me therefore begin at the last stage, and from there work through the model backwards until time 0.

3.2.1 Entry Decision of the Second Firm

Let me assume that duopoly profits only depend on the achieved technological parameters (γ^i, γ^j). Let the technology levels be γ^i for firm i, and γ^j for firm j. The expected future profits for firm j after both firms entered, discounted to the time of entry of the second firm, are defined as

$$D^j(\gamma^i, \gamma^j) := \int_0^\infty e^{-rt}\pi(\gamma^i, \gamma^j)dt. \tag{3.2}$$

I first characterize the entry decision of the second firm. Suppose firm i is already in the market and realized γ^i. Let the flow cost of research on that stage be ρ_j.[8]

[6] The first mover would experience speeded up depreciation of his specific capital if he undertakes research to replace himself. In such a situation it may be optimal to invest nothing.

[7] A strategy combination forms a subgame perfect equilibrium if it induces an equilibrium in any subgame of the original game.

[8] Here I introduce a research cost parameter ρ_j which may be lower than c_j. This will be used later to analyze spillover effects. One might also use the efficiency parameter λ_j for the comparative statics analysis of spillover effects. This does not change the qualitative results of the model.

Firm j now solves a control problem similar to that of the single firm in Chapter 2.2.2, i.e. it chooses an optimal threshold quality $\bar{\gamma}^j = \bar{\gamma}^j(\gamma^i)$,

$$\max_{\bar{\gamma}^j} \frac{-\rho_j + \lambda_j \int_{\bar{\gamma}^j}^{\infty} D^j(\gamma^i, \gamma^j) dF(\gamma^j)}{r + \lambda_j(1 - F(\bar{\gamma}^j))}. \tag{3.3}$$

The necessary condition for (3.3) is given by

$$rD^j(\gamma^i, \bar{\gamma}^j) = \lambda_j \int_{\bar{\gamma}^j}^{\infty} (D^j(\gamma^i, \gamma^j) - D^j(\gamma^i, \bar{\gamma}^j)) dF(\gamma^j) - \rho_j. \tag{3.4}$$

To derive further results let me impose the following assumption:

Assumption 3.3 *The expected profits $D^j(\gamma^i, \gamma^j)$ of the second entrant (i.e. firm j) are monotone increasing (but bounded) in its quality level.*

With this assumption I can state:

Lemma 3.1 *Let Assumption 3.3 hold. Then (3.4) is also sufficient for the solution to (3.3).*

Proof: The sufficient condition for (3.3) is given by

$$(r + \lambda_j(1 - F(\bar{\gamma}^j))) \frac{\partial D^j(\bar{\gamma}^j, \gamma^i))}{\partial \bar{\gamma}^j} > 0.$$

This is true from Assumption 3.3. □

Interpretation: Suppose first that $D^j(\gamma^i, \gamma^j)$ is monotonically decreasing in γ^j. In this case the second entrant would never continue investing after the first firm has entered. The resulting game resembles the case discussed in Lippman and Mamer [1993]. Second, let $D^j(\gamma^i, \gamma^j)$ be increasing up to a certain quality $\hat{\gamma}^j$ and decreasing thereafter. The second entrant would never continue investing beyond the point in time where it reached this turning point. In this case, assumption 3.3 contains no restriction whenever firm j is free to market a lower quality than its currently best. Let me coin this condition *free disposal of quality*. Under this condition $D^j(\gamma^i, \gamma^j)$ is weakly increasing whenever it is not montonically decreasing over the whole range. The boundedness assumption is necessary to guarantee a well-defined decision problem of the second entrant.

Variable Intensity

One may feel uncomfortable about the assumption that firms spend on R&D at a constant rate. I therefore shortly analyze the case of variable intensity, too. Assume that the firms can choose their intensity freely. To undertake research at an intensity λ, a convex flow cost $c_j(\lambda)$ has to be incurred (for firm j). The second entry problem of firm j then becomes

$$\max_{\lambda,\bar{\gamma}^j} \frac{-c_j(\lambda) + \lambda \int_{\bar{\gamma}^j}^{\infty} D^j(\gamma^i,\gamma^j)dF(\gamma^j)}{r + \lambda(1 - F(\bar{\gamma}^j))}. \tag{3.5}$$

For a given threshold strategy $\bar{\gamma}^j$ the optimal intensity solves

$$\begin{aligned} c_j'(\lambda)(r + \lambda(1 - F(\bar{\gamma}))) &= (r + \lambda(1 - F(\bar{\gamma})) \int_{\bar{\gamma}^j}^{\infty} D^j(\gamma^i,\gamma^j)dF(\gamma^j) \quad (3.6)\\ &+ (1 - F(\bar{\gamma}))(-c_j(\lambda) + \lambda \int_{\bar{\gamma}^j}^{\infty} D^j(\gamma^i,\gamma^j)dF(\gamma^j)). \end{aligned}$$

Together with the first order condition for an optimal $\bar{\gamma}$ one obtains

$$c_j'(\lambda) = \int_{\bar{\gamma}^j}^{\infty} (D^j(\gamma^i,\gamma^j) - D^j(\gamma^i,\bar{\gamma}^j))dF(\gamma^j). \tag{3.7}$$

Let me state the following proposition:

Proposition 3.1 *Suppose that firms can freely choose their optimal intensity. Let the instantaneous R&D cost function be convex ($c''(\lambda) > 0$). Then the optimal intensity is a decreasing function of the optimal threshold.*

Proof: Totally differentiate (3.7) to obtain

$$\frac{d\lambda}{d\bar{\gamma}^j} = -\frac{1 - F(\bar{\gamma}^j)}{c''(\lambda)}(W^j(\bar{\gamma}))' < 0.$$

This proves the claim.□

Discussion: With the previous result it is clear that one does not lose much generality if the analysis is restricted to the fixed intensity/fixed spendings case. Even when spendings may vary, firms would choose a constant rate. Nevertheless, for the following comparative statics one should keep in mind that a move of the optimal threshold in one direction causes a move of optimal intensity into the other direction in the variable intensity model.

Comparative Statics

Let Assumption 3.3 hold. From the discussion in Chapter 2.2.2 it is clear that the quality chosen by firm j negatively depends on research costs ρ_j and the discount rate r, while it depends positively on its research efficiency parameter λ_j.

Let me turn to the following question: How does the quality choice of the second entrant depend on the quality the first firm already employed? Suppose that the two firms try two develop vertically differentiated products. In this context the question is, how does the degree of differentiation between the two firms depend on the quality the first firm has introduced into the market. A similar question arises in the context of process innovations: How does a higher technological standard supplied by the first mover affect the technological standard the second firm develops?

To answer this question I totally differentiate equation (3.4):

$$\frac{\mathrm{d}\gamma^j}{\mathrm{d}\gamma^i} = \frac{-r(\frac{\partial D^j(\gamma^i,\bar{\gamma}^j)}{\partial\gamma^i}) + \lambda_j \int_{\bar{\gamma}^j}^{\infty}(\frac{\partial D^j(\gamma^i,\gamma^j)}{\partial\gamma^i} - \frac{\partial D^j(\gamma^i,\bar{\gamma}^j)}{\partial\gamma^i})dF(\gamma^j)}{(r+\lambda_j(1-F(\bar{\gamma}^j)))\frac{\partial D^j(\gamma^i,\bar{\gamma}^j)}{\partial\bar{\gamma}^j}}. \tag{3.8}$$

Note that the second entrant (firm j) chooses its quality threshold the lower, the higher the quality employed by the first entrant if equation (3.8) is negative. In addition, as long as (3.8) is smaller than one, the *quality gap* between the two firms decreases (given that the second entrant chooses a threshold above the first mover's quality) whenever the first firm enters with a higher quality.

The quality threshold of the second entrant is an increasing function of the quality employed by the first mover whenever

$$\lambda_j \int_{\bar{\gamma}^j}^{\infty}(\frac{\partial D^j(\gamma^i,\gamma^j)}{\partial\gamma^i} - \frac{\partial D^j(\gamma^i,\bar{\gamma}^j)}{\partial\gamma^i})dF(\gamma^j) > r(\frac{\partial D^j(\gamma^i,\bar{\gamma}^j)}{\partial\gamma^i}). \tag{3.9}$$

The left hand side of (3.9) is positive if

$$\frac{\partial D^j(\gamma^i,\gamma^j)}{\partial\gamma^i} > \frac{\partial D^j(\gamma^i,\bar{\gamma}^j)}{\partial\gamma^i},\ \forall\gamma^j > \bar{\gamma}^j, \tag{3.10}$$

while negative whenever the inequality is reversed.

Definition 3.1 *(i) The function $D^j(\gamma^i,\gamma^j)$ defined above exhibits* isotone differences *in (γ^i,γ^j) if $\forall\ \hat{\gamma}^i > \tilde{\gamma}^i$, and $\forall\ \hat{\gamma}^j > \tilde{\gamma}^j$*

$$D^j(\hat{\gamma}^i,\hat{\gamma}^j) + D^j(\tilde{\gamma}^i,\tilde{\gamma}^j) \geq D^j(\tilde{\gamma}^i,\hat{\gamma}^j) + D^j(\hat{\gamma}^i,\tilde{\gamma}^j).$$

(ii) Suppose that the payoff functions D^i, D^j exhibit isotone differences. The underlying game is a supermodular game.[9]

Interpretation: Every supermodular game is a game with *strategic complements.* The most prominent example of a supermodular game is the common Bertrand duopoly game with constant marginal costs. If one firm decreases (increases) its price, it is optimal for the rival to do the same. Another example is the standard model of price competition with vertically differentiated products with an uncovered market (assuming that consumers have linear demand functions that differ only in a taste parameter).

Assume that the order over the strategies of one player is reversed, i.e. the condition in Definition 3.1 is fulfilled for all $\hat{\gamma}^i < \tilde{\gamma}^i$. In this case the game is a game with *strategic substitutes* and the payoff function exhibits *antitone differences.* The classical example is the Cournot duopoly game with linear demand and constant marginal costs.

To apply the notion of supermodularity to the present problem, let $\tilde{\gamma}^i = \gamma^i$, $\hat{\gamma}^i = \gamma^i + h > \gamma^i$, $\tilde{\gamma}^j = \bar{\gamma}^j$ and $\hat{\gamma}^j = \gamma^j$. Supermodularity then implies that

$$D^j(\gamma^i + h, \gamma^j) + D^j(\gamma^i, \bar{\gamma}^j) \geq D^j(\gamma^i, \gamma^j) + D^j(\gamma^i + h, \bar{\gamma}^j) \tag{3.11}$$

or,

$$D^j(\gamma^i + h, \gamma^j) - D^j(\gamma^i, \gamma^j) \geq D^j(\gamma^i + h, \bar{\gamma}^j) - D^j(\gamma^i, \bar{\gamma}^j). \tag{3.12}$$

[9] The definition of a supermodular game here is only valid given the specification of the game above. In addition the payoff function of player i must be supermodular on the strategy set of player i, and the strategy sets must be complete lattices. These conditions are fulfilled here. For a more general discussion, see e.g Topkis [1979] and Vives [1990].

This in turn implies

$$\lim_{h\downarrow 0}\frac{D^j(\gamma^i+h,\gamma^j)-D^j(\gamma^i,\gamma^j)}{h}\geq \lim_{h\downarrow 0}\frac{D^j(\gamma^i+h,\bar{\gamma}^j)-D^j(\gamma^i,\bar{\gamma}^j)}{h} \tag{3.13}$$

and therefore also implies (3.10). Now suppose that the right hand side of (3.9) is negative. To give an intuition, this is the case if the firms develop differentiated products, competition is in prices in an uncovered market and the second entrant has a higher quality threshold as the quality of the first entrant. As noted above, at the same time this game is supermodular in (γ^i,γ^j). In this case I can state

Proposition 3.2 *Let the duopoly stage be a supermodular game. In addition, let Assumption 3.3 be fulfilled and let the right hand side of (3.9) be negative. Then the quality threshold of the second entrant is increasing in the quality of the first entrant.*

Remark: Note that the assumptions in the precedent proposition are sufficient, but by no means necessary. It would also be sufficient to assume that the second entrant's payoff function exhibit isotone differences.[10] The proposition holds also in the limiting case where (3.9) is of either sign and r goes to zero. In this case I can state

Proposition 3.3 *Let the payoff function $D^j(\gamma^i,\bar{\gamma}^j)$ of the second entrant (firm j) exhibit antitone differences and let r tend to zero. In addition, let Assumption 3.3 be fulfilled. Then the quality threshold of the second entrant is decreasing in the quality of the first entrant.*

Interpretation: An example in this direction is Cournot competition in the market stage. Firms develop process innovations that lead to different (constant) marginal costs in final production. Again, the assumption that r tends to zero is sufficient but not necessary. It is sufficient to assume that r is small enough, so that the inequality in (3.9) is reversed.

It remains to show how the *quality gap* between the first and the second entrant depends on the quality of the first mover (given that the innovation of the second

[10]Note that this condition is fulfilled in a greater class of games with vertically differentiated products. It is then only necessary to assume that the second entrant produces higher quality. See Shaked and Sutton [1982] for details.

entrant improves upon the innovation of the first entrant). Note that from equation (3.8) this gap is decreasing whenever

$$-r\left(\frac{\partial D^j(\gamma^i,\bar{\gamma}^j)}{\partial \gamma^i}\right) + \lambda_j \int_{\bar{\gamma}^j}^{\infty}\left(\frac{\partial D^j(\gamma^i,\gamma^j)}{\partial \gamma^i} - \frac{\partial D^j(\gamma^i,\bar{\gamma}^j)}{\partial \gamma^i}\right)dF(\gamma^j) < (r+\lambda_j(1-F(\bar{\gamma}^j)))\frac{\partial D^j(\gamma^i,\bar{\gamma}^j)}{\partial \bar{\gamma}^j}. \tag{3.14}$$

This can be rewritten as

$$\int_{\bar{\gamma}^j}^{\infty}\frac{\partial D^j(\gamma^i,\gamma^j)}{\partial \gamma^i}dF(\gamma^j) < ((r+\lambda_j(1-F(\bar{\gamma}^j)))\left[\frac{\partial D^j(\gamma^i,\bar{\gamma}^j)}{\partial \bar{\gamma}^j} + \frac{\partial D^j(\gamma^i,\bar{\gamma}^j)}{\partial \gamma^i}\right]. \tag{3.15}$$

Assume that an increase in the quality of the first entrant reduces the expected profits of the second entrant, and that the size of the direct effect from a marginal change in own quality always exceeds the size of the indirect effect from a marginal change in rival's quality, i.e.

$$\frac{\partial D^j(\gamma^i,\bar{\gamma}^j)}{\partial \bar{\gamma}^j} + \frac{\partial D^j(\gamma^i,\bar{\gamma}^j)}{\partial \gamma^i} > 0.$$

In this case the left hand side of (3.15) is negative, while the right hand side is positive. Therefore, the inequality is fulfilled. I can state

Proposition 3.4 *Let Assumption 3.3 hold. Assume that the expected profits of the second entrant decrease from a marginal increase in the quality of the first entrant, but let this effect be smaller in size than the direct effect from a marginal increase in own quality. In addition, assume that the second entrant chooses a higher quality than the first entrant. Then the quality gap is decreasing in the quality employed by the first entrant.*

Summary: Before I turn to the analysis of the first stage of the game, let me shortly summarize the results obtained so far. Let me concentrate on the case where the second entrant tries to develop a higher quality than the first entrant.

- The quality gap between the first and the second entrant depends on the quality of the first entrant and on the specification of the post-innovation market game.

- Quality levels are ***strategic complements*** in the market stage (i.e. the case of isotone differences). In this case a higher quality of the first entrant leads to a higher quality threshold of the second entrant. An example for this case is Bertrand competition with vertically differentiated products.

- Quality levels are ***strategic substitutes*** in the market stage (i.e. the case of antitone differences) and the discount rate r is small. In this case a higher quality of the first entrant leads to a lower quality threshold of the second entrant. An example is Cournot competition.

- Independent of the specification of the game in the market stage the quality gap between the two firms tends to decrease in the quality of the first entrant.

Especially the third point has an interesting implication. In some circumstances (i.e. Cournot competition on the product market) a very advanced innovation by an early innovator may hamper the subsequent technological evolution in an industry.

3.2.2 The First-Entry Game

Before the second firm enters, the first firm enjoys a temporary monopoly position. Instantaneous monopoly profits depend only on the quality the monopolist is using or bringing to market. Let them be denoted $\pi(\gamma^i)$, $i = 1, 2$. Before any firm has decided to implement its currently best innovation, both firms invest in R&D. This is reflected in Assumption 3.2.

Without loss of generality let me concentrate on firm i.[11] Suppose firm i realized a quality level γ^i. It has to decide whether to continue research to eventually find a better quality, or to introduce the current innovation into the market. If it decides to enter, while the other firm j has not entered yet, the present value of its expected future profits is calculated as follows.

First Entry Profits

During a certain time span firm i will be a monopolist in the industry, earning a flow profit $\pi(\gamma^i)$. Depending on the solution of the optimal control problem of firm j in the last section, firm j eventually enters at some time T_j. This variable

[11]The case of the other firm is symmetric.

is exponentially distributed with parameter $\lambda_j(1 - F(\gamma^j(\gamma^i)))$, where $\gamma^j(\gamma^i)$ is the optimal solution of the control problem according to (3.4). Thus the expected profit of firm i conditional on being first is given by

$$W^i(\gamma^i) := E\left[\frac{\pi(\gamma^i) + \lambda_j \int_{\bar{\gamma}^j(\gamma^i)}^{\infty} D^i(\gamma^i,\gamma^j)f(\gamma^j)d(\gamma^j)}{r + \lambda_j(1 - F(\bar{\gamma}^j(\gamma^i)))}\right]. \tag{3.16}$$

Second Entry Profits

Suppose on the other hand that firm i is preempted by firm j. In this case firm j is the first firm in the market. Suppose firm j entered with a quality level γ^j. Firm i then chooses its optimal response given that it is second. Let the optimal solution for firm i according to (3.4) be denoted $\bar{\gamma}^i(\gamma^j)$. The time that elapses until firm i for the first time realizes an innovation with a quality of at least $\bar{\gamma}^i(\gamma^j)$ is again exponentially distributed with parameter $\lambda_i(1 - F(\bar{\gamma}^i(\gamma^j)))$. Firm i spends R&D at a rate ρ_i. Its expected profits conditional on being second is thus given by

$$L^i(\gamma^j) := E\left[\frac{-\rho_i + \lambda_i \int_{\bar{\gamma}^i(\gamma^j)}^{\infty} D^i(\gamma^j,\gamma^i)f(\gamma^i)d(\gamma^i)}{r + \lambda_i(1 - F(\bar{\gamma}^i(\gamma^j)))}\right]. \tag{3.17}$$

Equilibrium of the First Entry Game

Until first entry, research costs are fixed at a rate c_i. Thus the expected profits of firm i, $i = 1,2$ are given by equation (3.18). Profits are maximized by the choice of a threshold strategy $\bar{\gamma}^i$ for firm $i = 1,2$:

$$\max_{\bar{\gamma}^i} \frac{-c_i + \lambda_i \int_{\bar{\gamma}^i}^{\infty} W^i(\gamma^i)f(\gamma^i)d\gamma^i + \lambda_j \int_{\bar{\gamma}^j}^{\infty} L^i(\gamma^j)f(\gamma^j)d(\gamma^j)}{r + \sum_{i=1}^{2} \lambda_i(1 - F(\bar{\gamma}^i))}. \tag{3.18}$$

As in the previous section, I derive the necessary condition to (3.18). This condition is given by

$$\begin{aligned} rW^i(\bar{\gamma}^i) = -c_i \; &+ \; \lambda_i \int_{\bar{\gamma}^i}^{\infty} [W^i(\gamma^i) - W^i(\bar{\gamma}^i)]f(\gamma^i)d\gamma^i \\ &+ \; \lambda_j \int_{\bar{\gamma}^j}^{\infty} [L^i(\gamma^j) - W^i(\bar{\gamma}^i)]f(\gamma^j)d\gamma^j. \end{aligned} \tag{3.19}$$

Lemma 3.2 *A sufficient condition for a best response threshold for firm i to $\bar{\gamma}^j$ is*

$$(r + \sum_{i=1,2} \lambda_i(1 - F(\bar{\gamma}^i)))(W^i(\bar{\gamma}^i))' > 0. \tag{3.20}$$

Therefore I need

Assumption 3.4 *Let the first mover expected profit be an increasing (but bounded) function in the quality of the first mover.*

Discussion: Under Assumption 3.4 condition (3.20) is always fulfilled. Note that Assumption 3.4 is not restrictive if *free disposal of quality* is possible (see the discussion on Assumption 3.3 above). The boundedness assumption is sufficient to guarantee that expectations are always well-defined, and that no firm will find it profitable to continue investing in R&D forever. This defines an implicit upper bound on the threshold that will be chosen in equilibrium.

Back to equation (3.19). Its left hand side is the average future profit rate the first entrant expects if he enters with his threshold quality. The right hand side represents the expected profit change from investing an amount of time dt longer which involves a cost $c_i dt$. With an instantaneous probability of $\lambda_i(1 - F(\bar{\gamma}^i))dt$ firm i will make the next innovation with a higher quality than the threshold. This results into an expected profit of

$$\frac{\int_{\bar{\gamma}^i}^{\infty} W^i(\gamma^i) f(\gamma^i) d\gamma^i}{(1 - F(\bar{\gamma}^i))}. \tag{3.21}$$

This opportunity is reflected in the first integral. The second integral is responsible for the *pre-emption* or *interaction* effect. If firm i waits for another infinitesimal amount of time dt, firm j succeeds in completing an innovation above its threshold with instantaneous probability $\lambda_j(1 - F(\bar{\gamma}^j))dt$. In this case firm i loses the first mover profit $W^i(\bar{\gamma}^i)$, and receives the expected second mover profit in turn, which is given by

$$\frac{\int_{\bar{\gamma}^j}^{\infty} L^i(\gamma^j) f(\gamma^j) d\gamma^j}{(1 - F(\bar{\gamma}^j))}. \tag{3.22}$$

I call this second integral in (3.19) the *interaction term*. I can now state the first result:

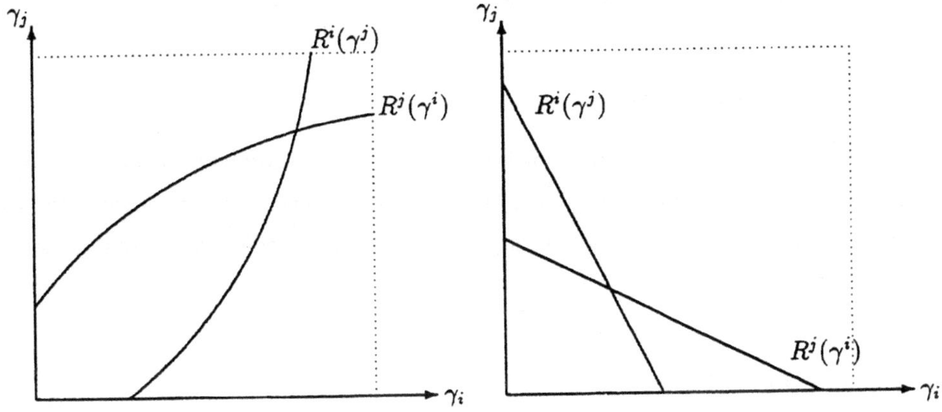

Figure 3.2 (a)

Figure 3.2 (b)

Figure 3.2: First- and Second-Mover Advantages

Proposition 3.5 *Let Assumptions 3.2 and 3.4 hold. (i) Firms choose threshold strategies to determine their stopping time in the first-entry game. (ii) The best response of firm i, $\bar{\gamma}^i$, is increasing in $\bar{\gamma}^j$, if and only if $W^i(\bar{\gamma}^i)$ exceeds $L^i(\bar{\gamma}^j)$, and is decreasing otherwise.*

Proof:(i) The first part is Lemma 1 in Lippman and Mamer [1993], p. 110f. (ii) Total differentiation of equation (3.19) yields

$$\frac{d\bar{\gamma}^i}{d\bar{\gamma}^j} = \frac{\lambda_j f(\bar{\gamma}^j)[W^i(\bar{\gamma}^i) - L^i(\bar{\gamma}^j)]}{(r + \sum_{i=1,2} \lambda_i(1 - F(\bar{\gamma}^i)))(W^i(\bar{\gamma}^i))'}.$$

Note that the denominator is always positive from Assumption 3.4. Thus, if $W^i > L^i$, an increase (decrease) in $\bar{\gamma}^j$ leads to an increase (decrease) in $\bar{\gamma}^i$. □

The connection between *first-mover advantages* and *increasing best responses* as well as between *second-mover advantages* and *decreasing best responses* are depicted in Figure 3.2. With Proposition 3.5 I can show the following theorem for the case of two firms:

Theorem 3.1 *Let Assumptions 3.1–3.4 be fulfilled. Then there exists a subgame perfect equilibrium for the sequential entry game with two firms.*

Proof: See Appendix (A.1).

The generalization to the case of n firms does not pose a problem in the case where there are first-mover advantages.[12] In the case of second-mover advantages this is more problematic, because the algebraic structure used here to show existence cannot be used directly to show existence in the case of n firms and second-mover advantages.

Variable Intensity

As in the second entry stage let me shortly derive the optimal R&D intensity in the first entry game. The firm solves

$$\max_{\bar{\gamma}^i, \lambda^i} \frac{-c_i(\lambda^i) + \lambda^i \int_{\bar{\gamma}^i}^{\infty} W^i(\gamma^i) f(\gamma^i) d\gamma^i + \lambda^j \int_{\bar{\gamma}^j}^{\infty} L^i(\gamma^j) f(\gamma^j) d(\gamma^j)}{r + \sum_{i=1}^{2} \lambda^i (1 - F(\bar{\gamma}^i))}. \tag{3.23}$$

The same reasoning as before gives the first order condition for the optimal R&D-intensity in the first entry game:

$$c_i'(\lambda^i) = \int_{\bar{\gamma}^i}^{\infty} (W(\gamma) - W(\bar{\gamma}^i)) dF(\gamma). \tag{3.24}$$

I can thus state

---

[12]Consider the symmetric case of n instead of two identical firms in the case of first mover advantages. The first order condition in this case becomes (indices are suppressed)

$$(r + n(1 - F(\bar{\gamma}))W(\bar{\gamma}) = -c + \lambda \int_{\bar{\gamma}} W(\gamma) dF(\gamma) + (n-1)\lambda \int_{\bar{\gamma}} L(\gamma) dF(\gamma).$$

Here I used a result that is derived below (Theorem 3.2), namely that first-mover advantages imply symmetric equilibria. Total differentiation shows that the equilibrium threshold is decreasing in the number of firms because of first-mover advantages:

$$\frac{d\bar{\gamma}}{dn} = \frac{\lambda \int_{\bar{\gamma}} (L(\gamma) - W(\gamma)) dF(\gamma)}{(r + n\lambda(1 - F(\bar{\gamma})))W'(\bar{\gamma})} < 0.$$

Proposition 3.6 *Suppose that firms can freely choose their R&D-intensity. The instantaneous R&D cost function is convex* ($c_i''(\lambda^i) > 0$). *(i) The optimal R&D-intensity does not depend on rival action. (ii) It is also independent from expected second-mover profits. (iii) It is a decreasing function of the firm's own quality threshold.*

Proof: Follows from equation (3.24) and Proposition 3.1.

Remark: The restriction to the case of fixed intensity and fixed spendings can be justified by the previous proposition. Keep in mind for the comparative statics that the optimal intensity in the variable intensity model moves in the opposite direction from parameter changes than the optimal threshold of a firm.

An Example

Consider an example with two identical firms. Suppose that there is no demand for an innovation of quality lower than some $\hat{\gamma}$. Furthermore, for $\gamma \geq \hat{\gamma}$ consumers can only distinguish between innovations in an interval $[\hat{\gamma}, \tilde{\gamma})$, $\hat{\gamma} < \tilde{\gamma}$ and innovations with quality higher than $\tilde{\gamma}$. Thus, the expected profit functions of the firms are step functions with jumps at $\hat{\gamma}$ and $\tilde{\gamma}$. Between those values the profit functions are flat. Starting research is nevertheless assumed to be profitable.

Consequently, the analysis can be restricted to the choice of thresholds from the set $\{\hat{\gamma}, \tilde{\gamma}\}$ in the first entry game (firms naturally will not choose intermediate values as threshold, since a higher threshold implies stochastically higher research expenditures while profit functions are flat). Assume in addition that the firm that is the second entrant always finds it profitable to choose the higher threshold, irrespectively of the quality the first firm produces. Therefore I can further restrict the analysis to the first entry game.

The resulting payoff matrix is as follows:

1\2	$\hat{\gamma}$	$\tilde{\gamma}$
$\hat{\gamma}$	$V^1(\hat{\gamma}, \hat{\gamma}), V^2(\hat{\gamma}, \hat{\gamma})$	$V^1(\hat{\gamma}, \tilde{\gamma}), V^2(\hat{\gamma}, \tilde{\gamma})$
$\tilde{\gamma}$	$V^1(\tilde{\gamma}, \hat{\gamma}), V^2(\tilde{\gamma}, \hat{\gamma})$	$V^1(\tilde{\gamma}, \tilde{\gamma}), V^2(\tilde{\gamma}, \tilde{\gamma})$

The entries into the payoff matrix are calculated according to equation (3.18). I derive the pure strategy equilibria of the game. Let me concentrate on firm 1. The case of firm 2 is symmetric.

Suppose that firm 2 chooses the strategy $\hat{\gamma}$. Firm 1 prefers $\hat{\gamma}$ to $\tilde{\gamma}$ if $V^1(\hat{\gamma},\hat{\gamma}) > V^1(\tilde{\gamma},\hat{\gamma})$. This is the case if

$$\begin{aligned} rW^1(\hat{\gamma}) &> -c+\lambda((1-F(\tilde{\gamma}))(W^1(\tilde{\gamma})-W^1(\hat{\gamma})+ \\ & \quad L^1(\tilde{\gamma})-L^1(\hat{\gamma}))+(F(\tilde{\gamma})-F(\hat{\gamma}))(L^1(\hat{\gamma})-W^1(\hat{\gamma}))). \end{aligned} \tag{3.25}$$

Suppose instead that firm 2 chooses the strategy $\tilde{\gamma}$. Firm 1 plays also $\tilde{\gamma}$ if $V^1(\tilde{\gamma},\tilde{\gamma}) > V^1(\hat{\gamma},\tilde{\gamma})$. This is the case if

$$rW^1(\hat{\gamma}) < -c+\lambda((1-F(\tilde{\gamma}))(W^1(\tilde{\gamma})-W^1(\hat{\gamma})+L^1(\tilde{\gamma})-L^1(\hat{\gamma}))). \tag{3.26}$$

Together with the symmetric conditions for firm 2 this determines the cases in which there may be symmetric equilibria. If both inequalities in the equations (3.25),(3.26) hold there are two symmetric equilibria. If only one of the two inequalities is reversed one obtains one symmetric equilibrium. If both are reversed, only asymmetric equilibria exist. I can state the following characterizing result:

Proposition 3.7 *Consider the game in the present example. The following statements are true:*

(i) There are two symmetric equilibria $\Longrightarrow$ $W^i(\hat{\gamma}) > L^i(\hat{\gamma})$.

(ii) There is no symmetric equilibrium $\Longrightarrow$ $W^i(\hat{\gamma}) < L^i(\hat{\gamma})$.

(iii) $(\hat{\gamma},\hat{\gamma})$ *is an equilibrium with* $W^i(\hat{\gamma}) < L^i(\hat{\gamma})$. *Then this equilibrium is unique.*

Proof: (i) Substract (3.26) from (3.25). (ii) follows directly from (i). (iii) Show that in case (3.25) holds and $W^i(\hat{\gamma}) < L^i(\hat{\gamma})$ it follows that (3.26) does not hold.□

Interpretation: In a game with *second-mover advantages* ($L^i > W^i$) no firm wants to be the first to enter the market. In the review of the literature on models of sequential entry I referred to the example of a game with second-mover advantages provided by Dasgupta [1986]. Remember the game in Figure 2.4 (Chapter 2.4).

A\B	enter now	enter later
enter now	(4, 4)	(5, 10)
enter later	(10, 5)	(8, 8)

This is the example of a deterministic *waiting game* with a second-mover advantage. How does this game correspond to the example above?

In a deterministic game firms choose their entry times (as pure strategies), while in a stochastic game the optimal strategies chosen by firms induce a probability distribution over entry times. It is therefore possible that a firm that has chosen a higher threshold strategy in an asymmetric equilibrium of the stochastic game nevertheless becomes the first active firm in the market.

There is a close connection between the two games. As stated in the Proposition above, two symmetric equilibria imply first-mover advantages. The reverse, however, is not true. There might be an equilibrium with second-mover advantages that is symmetric. In the example this implies that the equilibrium is then unique. The result should be interpreted carefully, since it is an artefact of the discretization of the strategy space. It does not carry over to the general case as I will show later in this chapter. Nevertheless, whenever there is a symmetric equilibrium with second-mover advantages, this equilibrium is the unique equilibrium with this feature in the generalized setting.

Another result that carries over is that first-mover advantages cannot prevail in an asymmetric equilibrium. Whenever an equilibrium in the first entry game implies different strategies chosen by the firms the resulting characteristics entail second-mover advantages.

3.3 Comparative Statics

Let Assumptions 3.1–3.4 hold.[13]

Furthermore, let the equilibrium be unique.[14] I distinguish two cases, which are depicted in Figure 3.2.[15] If first-mover expected profits are higher than second-mover expected profits the first-entry strategies of the firms are strategic complements (as in Figure 3.2 (a)). If the reverse is true, the first-entry game is a game with strategic substitutes (as in Figure 3.2 (b)).

[13] A more detailed discussion of the underlying algebraic tools used in the following section is deferred to Appendix (A.1) to guarantee readability.

[14] The consequences of multiple equilibria are discussed below.

[15] The geometrical demonstrations in this section are qualitative and serve only as a clarifying device. The exact functional form depends on the specification of the underlying game.

3.3.1 First-Mover Advantages: The Case of Strategic Complements

Consider first $W^i > L^i$, i.e. the threshold strategies of the two firms are *strategic complements.* If one firm increases its threshold the best response of the rival firm is to increase its threshold, too.

In a game with strategic complements, comparative statics results are easy to determine. If a change in the parameter set leads to a *ceteris paribus* increase in the threshold for one firm, the new equilibrium will show a higher threshold for both firms. Thus, for *qualitative comparative statics* it is sufficient to concentrate on the first order conditions for one firm, since the direct effects always have the same direction as higher order feedback effects in equilibrium. The comparative statics in the case where there are first-mover advantages are collected in Proposition 3.8. See also Figure 3.3 for a graphical exposition of the comparative statics results.

Proposition 3.8 *Let Assumptions 3.1–3.4 hold. Furthermore, let $W^i > L^i$, $i = 1, 2$ in the unique equilibrium. The equilibrium thresholds $(\bar{\gamma}^i, \bar{\gamma}^i)$ of both firms are decreasing in the research costs c_i and c_j as well as in the discount rate r. They are increasing in the efficiency parameters λ_i and λ_j.*

Proof: It suffices to verify the properties for the threshold of firm i. An increase in some parameter leeds to an increase of the equilibrium threshold if the profit function of the respective firm is increasing in the threshold and has isotone differences in that parameter and the threshold variable. This can be verified by totally differentiating the first order condition in (3.19). Total differentiation of (3.19) taking $\bar{\gamma}^i, c_i$ as variables yields

$$\frac{d\bar{\gamma}^i}{dc_i} = \frac{-1}{(r + \sum_{i=1}^2 \lambda_i(1 - F(\bar{\gamma}^i)))(W^i)'(\bar{\gamma}^i)} < 0,$$

since $(W^i)'(\bar{\gamma}^i)$ is positive. A change in the discount rate leads to

$$\frac{d\bar{\gamma}^i}{dr} = \frac{-W^i(\bar{\gamma}^i)}{(r + \sum_{i=1}^2 \lambda_i(1 - F(\bar{\gamma}^i)))(W^i)'(\bar{\gamma}^i)} < 0,$$

while a change in the efficiency parameter λ_i yields

$$\frac{d\bar{\gamma}^i}{d\lambda_i} = \frac{\int_{\bar{\gamma}^i}^{\infty} (W^i(\gamma^i) - W^i(\bar{\gamma}^i))f(\gamma^i)d\gamma^i}{(r + \sum_{i=1}^2 \lambda_i(1 - F(\bar{\gamma}^i)))(W^i)'(\bar{\gamma}^i)} > 0.$$

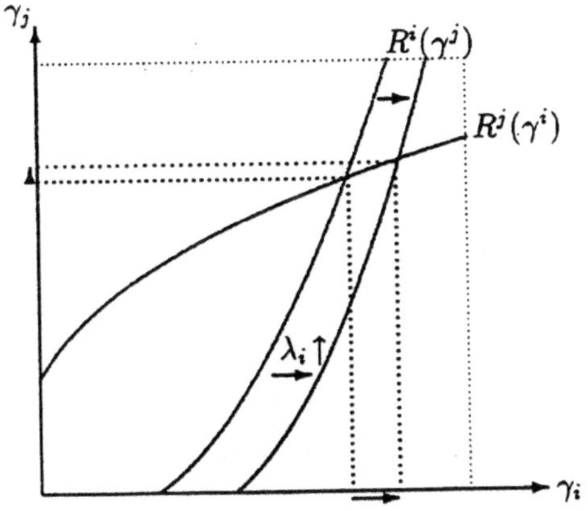

Figure 3.3 (a): Increase in λ_i

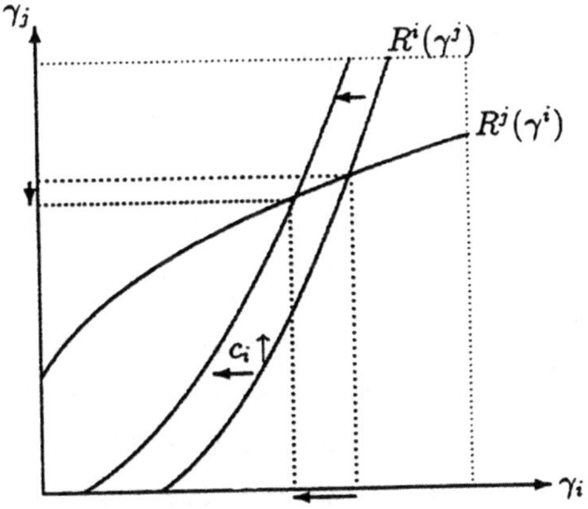

Figure 3.3 (b): Increase in c_i

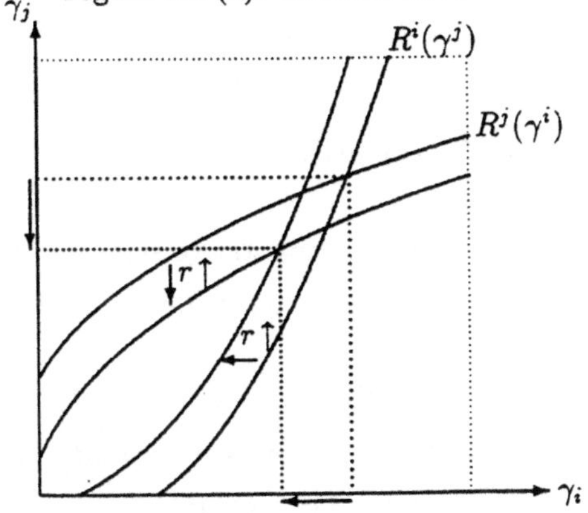

Figure 3.3 (c): Increase in r

Figure 3.3: Comparative Statics with First-Mover Advantages

Since the best response of the rival is increasing in $\bar{\gamma}^i$, the same qualitative comparative static results hold for firm j with respect to a parameter change for firm i.□

3.3.2 Second-Mover Advantages: The Case of Strategic Substitutes

Suppose now that $L^i > W^i$. This is the case where the quality levels of firms are *strategic substitutes*, i.e. an increase in the threshold of firm i leads to a decreased threshold of firm j as a best response. To analyze this case, I propose the following reformulation of the game. The strategy set of firm j is unchanged, but firm i chooses a threshold $\zeta^i \in (-\infty, 0]$ to maximize its profits, where $\zeta^i := -\gamma^i$.[16]

It is easy to verify that by this reformulation the game is again turned into a game with strategic complementarities with respect to (γ^j, ζ^i). I derive the qualitative comparative statics result in this new game and then translate the result back to the original game. Instead of (3.19) we have

$$rW^i(-\bar{\zeta}^i) = -c_i + \lambda_i \int_{-\bar{\zeta}^i}^{\infty} [W^i(-\zeta^i) - W^i(-\bar{\zeta}^i)] f(-\zeta^i) d\zeta^i \qquad (3.27)$$
$$+ \lambda_j \int_{\bar{\gamma}^j}^{\infty} [L^i(\gamma^j) - W^i(-\bar{\zeta}^i)] f(\gamma^j) d\gamma^j.$$

As in the previous case, I can now state the qualitative comparative statics of the game from (3.27) alone. A graphical illustration is given in Figure 3.4.

Proposition 3.9 *Let Assumptions 3.1–3.4 hold. Furthermore, let $W^i < L^i$, $i = 1, 2$ in the unique equilibrium. The threshold $\bar{\gamma}^i$ of firm i increases in own efficiency λ_i and in the research costs c_j of its rival. It decreases in the discount rate r, in the own research costs c_i as well as in the efficiency λ_j of its rival.*

Proof: I will again look at firm i. Totally differentiating (3.27) with respect to (ζ^i, c_i) yields

$$\frac{d\bar{\zeta}^i}{dc_i} = \frac{-1}{(r + \sum_{i=1}^{2} \lambda_i (1 - F(-\bar{\zeta}^i)))(\partial W^i(-\bar{\zeta}^i)/\partial \zeta^i)} > 0,$$

[16] See Chapter 3.2.1 above for a discussion of *order reversion*.

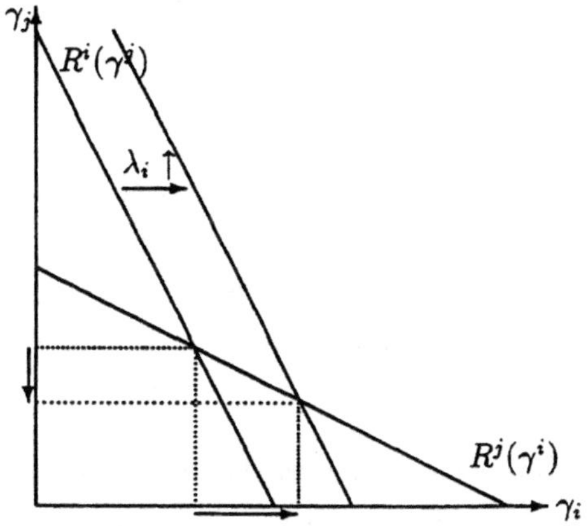

Figure 3.4 (a): Increase in λ_i

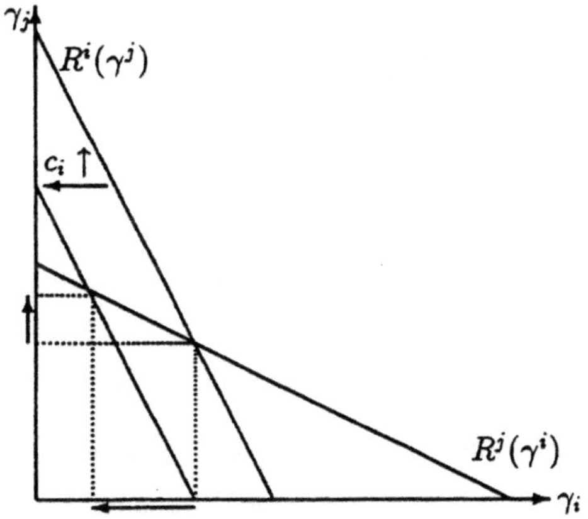

Figure 3.4 (b): Increase in c_i

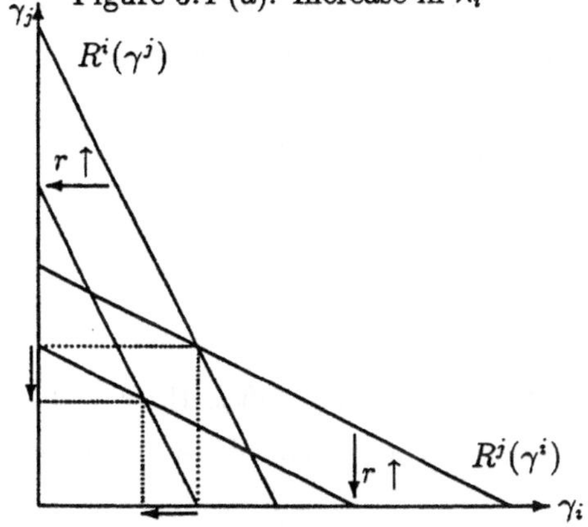

Figure 3.4 (c): Increase in r

Figure 3.4: Comparative Statics with Second-Mover Advantages since this time $\partial W^i(-\bar{\zeta}^i)/\partial\zeta^i < 0$. For the dependence on r we have

$$\frac{\mathrm{d}\bar{\zeta}^i}{\mathrm{d}r} = \frac{-W^i(-\bar{\zeta}^i)}{(r+\sum_{i=1}^2 \lambda_i(1-F(-\bar{\zeta}^i)))(\partial W^i(-\bar{\zeta}^i)/\partial\zeta^i)} > 0,$$

while a change in efficiency λ_i yields

$$\frac{\mathrm{d}\bar{\zeta}^i}{\mathrm{d}\lambda_i} = \frac{\lambda_i \int_{-\bar{\zeta}^i}^{\infty} [W^i(-\zeta^i) - W^i(-\bar{\zeta}^i)] f(-\zeta^i) d\zeta^i}{(r+\sum_{i=1}^2 \lambda_i(1-F(-\bar{\zeta}^i)))(\partial W^i(-\bar{\zeta}^i)/\partial\zeta^i)} < 0.$$

Since (ζ^i, γ^j) are strategic complements, an increase in the research costs c_i for firm i leads to an increase in the optimal threshold $\bar{\gamma}^j$ of firm j, while an increase in λ_i has the opposite effect on $\bar{\gamma}^j$. A change in r leads to a change in the same direction for $\bar{\zeta}^i$, and to a change in the opposite direction for $\bar{\gamma}^j$. Substituting back $\gamma^i = -\zeta^i$ leads to the remaining assertions of the proposition . □

Remark: In the case with second-mover advantages I reversed the order over the strategy set of one player to generate a game with strategic complementarities. Note that this is not always possible if there are more than two firms. The algebraic fixed-point theorem used in the proof of Theorem 3.1 is only applicable to games with strategic complementarities. The generalization to more than two players thus poses non-trivial problems for the existence of equilibrium.

3.3.3 Multiple Equilibria

So far, I concentrated on games with a unique equilibrium. Often there will be multiple equilibria in threshold strategies for the first entry game. What can be said in that case?

Suppose that there are multiple equilibria. Before I derive comparative statics results for the whole set of equilibria, let me first prove that all equilibria are symmetric in the case with first mover advantages. Furthermore, if there exists an equilibrium with second-mover advantages it is the only symmetric equilibrium with this property. For simplicity, I consider the symmetric game only. In addition, assume that second-mover profits are decreasing in the quality of the first entrant.[17]

[17] See Chapter 3.2.1 for a discussion.

Theorem 3.2 *(i) Suppose that $W > L$, and the two firms are identical. Only symmetric equilibria exist. (ii) Suppose that there exists a symmetric equilibrium $\bar{\gamma}$ with $W(\bar{\gamma}) < L(\bar{\gamma})$. Then there is no other symmetric equilibrium $\hat{\gamma}$ with this property.*

Proof: See Appendix A.2

Remark: There may be asymmetric equilibria in the game with second-mover advantages. The positive result that there exist only symmetric equilibria depends crucially on the fact that there are first-mover advantages.

Discussion: Dutta and Rustichini [1993] provide an example of a game with second-mover advantages where only asymmetric equilibria exist.[18] I show that there may also be a symmetric equilibrium with second-mover advantages. Their result is not directly comparable to the present model, where there may be asymmetric equilibria in the first-entry game. The reason is that in their model the innovation process is unidimensional and the roles of the firms as first or second mover are "predetermined". The asymmetry in their model reflects the difference in entry times between the leader and follower, whereas in my model it may well be that not only the follower chooses a later entry time than the leader, but that the firms may differ in their first-entry strategies as well.

The difference between the present approach and Dutta and Rustichini is pretty much the same as in the example presented in the last section. Dutta and Rustichini allow for stochastic improvement of technology, but this stochastic process is not influenced by the firms and it is the same for both firms. Thus, the difference in the technological standard available to the firms is non-random, while it is a random variable in my model.

No matter whether equilibria are symmetric or not, they can be ranked according to the order of the threshold strategies for the firms. A *higher* strategy for player i is a higher quality threshold $\bar{\gamma}^i$ in the game with first-mover advantages. In the game with second-mover advantages the order for one of the players is reversed (See the comparative statics exercise and Chapter 3.2.1 above). In this sense, a equilibrium is *higher* than another equilibrium in the product order over the joint strategy space if both firms choose a higher threshold in the former equilibrium with at least one

[18]See Chapter 3.4 below for a description of their model.

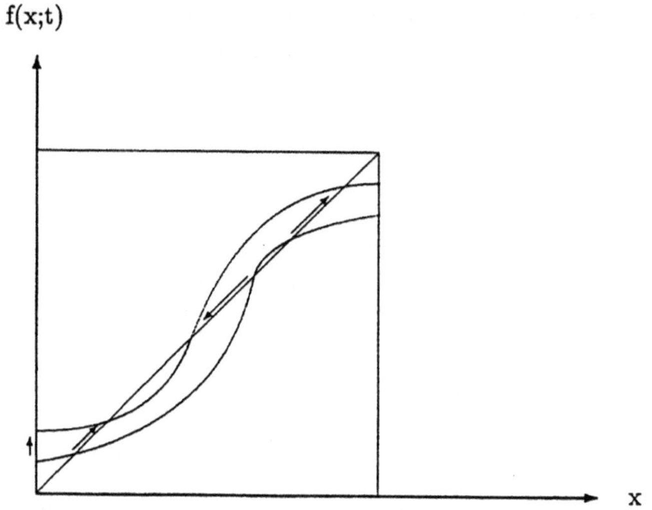

Figure 3.5: Monotone Selection of Fixed Points

strict inequality. Since both games show strategic complementarities with respect to the chosen order, all equilibria can be ranked, beginning with the lowest equilibrium, where both firms choose the lowest threshold strategies up to the highest. To fix the comparative static results derived so far for the case of multiple equilibria, I state the next proposition.

Proposition 3.10 *Suppose that the first-entry game has multiple equilibria. The equilibria can be completely ranked with respect to the product order over the joint strategy space introduced before. There exists a highest and a lowest equilibrium, and the comparative statics results derived for the case with a unique equilibrium also hold for the highest and the lowest equilibrium in a game with multiple equilibria.*

Proof: See Appendix A.3

The comparative statics results with respect to the highest and lowest equilibria do not translate to the case of an arbitrary equilibrium. Intermediate equilibria can move discontinuously with parameters and may even change the direction of movement in different parameter regions. In Figure 3.5 the function $f(x;t)$ is depicted for two values of the parameter t. An increase in t shifts the function f up. The set of fixed points also shifts upwards, while the fixed point in the middle shifts downwards. A further increase would lead even to the disappearence of this fixed point. The monotone comparative statics result is only applicable to the selections of the smallest and the greatest fixed points.[19]

Nevertheless, Proposition 3.10 is a strong result. It states that the complete set of equilibria is moving monotonically with the parameters of the system. From a positive point of view this sort of result is an enormous improvement over comparative statics results for selected equilibria. As long as there exist no tractable and justifiable equilibrium selection theory, information about the behavior of the set of equilibria (if there are many) is extremely valuable. I think that this is one of the important strenghts of the approach taken for the analysis of the present chapter.

3.3.4 Spillover Effects and Technology Choice

To conclude this section, I will shortly discuss the influence of spillover effects after first entry. An extensive discussion of the implications of informational spillovers for rent protection and timing decisions is provided in the next chapter.

Suppose that the first firm that brings its innovation to the market provides a valuable source of information for the rival that is still in the research stage. This can be information about the specifications of the pioneer's innovation, or about market conditions and consumer tastes. These spillover effects may lead to increased efficiency of the researching firm. I represent such an effect by a lower flow cost $\rho_i < c_i$. Alternatively, a higher efficiency parameter λ_i might result.[20]

An increased efficiency of research after the first innovator entered the market has two effects. First, second-mover expected profits (L) increase. The second effect is that the late mover tends to lenghten its research stage and to enter with a more advanced quality.[21] This has the effect that the first innovator enjoys a longer time

[19] Here it is necessary to assume that the strategy space is a (conditionally) complete lattice, since otherwise it is not guaranteed that those selections exist. See Appendix A.3 for details.

[20] See for example Mansfield, Wagner and Schwartz [1981] for the empirical relevance. Development costs tend to be lower for second movers.

[21] See Chapter 3.2.1 above.

period of monopoly. Remember that the first mover expected profits are

$$W^i(\gamma^i) := E\left[\frac{\pi(\gamma^i) + \lambda_j \int_{\bar{\gamma}^j(\gamma^i)}^{\infty} D^i(\gamma^i,\gamma^j)dF(\gamma^j)}{r + \lambda_j(1 - F(\bar{\gamma}^j(\gamma^i)))}\right]. \tag{3.28}$$

Spillovers lead to a *ceteris paribus* increase of the threshold of the second mover. The effect on first mover's profits are

$$\frac{\partial W^i}{\partial \bar{\gamma}^j} = \frac{\pi(\gamma^i) - rD^i(\gamma^i,\bar{\gamma}^j) + \lambda_j \int_{\bar{\gamma}^j(\gamma^i)}^{\infty} (D^i(\gamma^i,\gamma^j) - D^i(\gamma^i,\bar{\gamma}^j))dF(\zeta^j)}{(r + \lambda_j(1 - F(\bar{\gamma}^j(\gamma^i))))^2}. \tag{3.29}$$

Interpretation: Note that $\pi(\gamma^i) - rD^i(\gamma^i,\bar{\gamma}^j) > 0$ for all reasonable examples of market competition. The term $D^i(\gamma^i,\gamma^j) - D^i(\gamma^i,\bar{\gamma}^j)$ for $\gamma^j \geq \bar{\gamma}^j$ can either be positive or negative. If the innovation is cost reducing, this second term is likely to be negative. In the case of price competition with product differentiation and $\bar{\gamma}^j > \gamma^i$ the the term can also be positive, since duopoly profits then increase in the quality gap between the firms. The effect is therefore indeterminate. Only in the case of product innovation and price competition spillovers are definitely beneficial for the first entrant. To summarize the discussion so far, let me state

Lemma 3.3 *Consider the R&D-cost reducing spillovers discussed above. An increase in the spillover rate (i) increases the expected profits of the second entrant. (ii) At the same time the second entrant chooses a higher threshold strategy. (iii) The effect on the profits of the first entrant are indeterminate: There is a positive effect, because entry of the second firm tends to be later. The higher threshold can lead either to an increase or decrease in first-mover profits, dependent on the form of competition on the market stage.*

Suppose the *ceteris paribus* effect on first-mover profits is positive. How does this affect the strategy choice in the first entry game? To keep the analysis simple, suppose that first-mover expected profit uniformly increases by k units.[22] Equation

[22] By *uniform increase* I mean that the profits of the first entrant increase by k units, irrespectively which quality he realizes.

(3.19) becomes

$$r(W^i(\bar{\gamma}^i) + k) = -c_i \; + \; \lambda_i \int_{\bar{\gamma}^i}^{\infty} [W^i(\gamma^i) - W^i(\bar{\gamma}^i)] f(\gamma^i) d\gamma^i \qquad (3.30)$$

$$+ \; \lambda_j \int_{\bar{\gamma}^j}^{\infty} [L^i(\gamma^j) - (W^i(\bar{\gamma}^i) + k)] f(\gamma^j) d\gamma^j .$$

As is easily seen, the threshold of firm i has to decrease.

A uniform increase of k units in expected second-mover profits has the reverse effect. This time equation (3.19) becomes

$$rW^i(\bar{\gamma}^i) = -c_i \; + \; \lambda_i \int_{\bar{\gamma}^i}^{\infty} [W^i(\gamma^i) - W^i(\bar{\gamma}^i)] f(\gamma^i) d\gamma^i \qquad (3.31)$$

$$+ \; \lambda_j \int_{\bar{\gamma}^j}^{\infty} [L^i(\gamma^j) + k - W^i(\bar{\gamma}^i)] f(\gamma^j) d\gamma^j .$$

Thus, we have

Proposition 3.11 *A uniform increase in first-mover profits for firm i leads to a lower first-entry threshold, while a uniform increase in second-mover profits leads to an increase in the threshold.*

Interpretation: Concerning spillover effects, no clear picture emerges. There are two counterveiling effects (at least in the case of product innovations). Due to an increase in second-mover profits, the first-entry thresholds should increase, but the tendency that also first-mover profits may increase points into the opposite direction. Note that this result is similar to the result obtained by Spence [1984].[23] He analyzed different degrees of spillovers and their implications for innovation incentives. The main message of his paper is that spillovers do not necessarily spoil innovation incentives. This view is also confirmed by the empirical literature on the subject. Levin *e.a.* [1987] state in their remarks on policy (p.816):

> "Stronger appropriability will not yield more innovation in all contexts and, where it does, innovation may come at excessive cost."

[23]See also Cohen and Levinthal [1989]. In addition to the analysis undertaken by Spence they emphasize the role of accommodating research for the absorption capacity of spillovers.

An interesting corollary to the last proposition is that a uniform shift from first-mover expected profits to second-mover expected profits leads to an increase in the quality threshold chosen by the firm even if overall expected profits do not change. This indicates how the choice of first-entry quality is shaped by relative first- and second-mover expected profits.

3.4 The Endogeneity of Competitive Structure: Mixed Games

In the preceding analysis I treated the two polar cases of *first-mover advantages* and *second-mover advantages.* As I have shown the relative advantages of early and late entrants determine the whole structure of the model (see Proposition 3.5). The reaction curve of firm i is upward-sloping in rival's quality threshold if $L^i(\bar{\gamma}^j) < W^i(\bar{\gamma}^i)$, and downward-sloping if the inequality is reversed. From Assumption 3.4, we have W^i increasing in its argument, while L^i is decreasing in $\bar{\gamma}^j$ whenever duopoly profits of the second entrant decrease in rival's quality. Since this is true in most standard models of competition, it is also assumed here.

Three Classes of Games

Actually, the strategic structure of the game is endogenously determined. In the present model three classes of games can be distinguished (of which the first two classes are the polar cases already discussed in previous sections).

- The *first* class contains the *games with first-mover advantages.* In this class profits accruing to the first innovator always exceed profits of late innovators. In equilibrium, the first-entry quality threshold of a firm is increasing in the quality threshold of its rival. The structure of the game is a *race*, and the first innovator enters the market with a relatively low quality.

- The *second* class contains the *games with second-mover advantages.* In this class the expected equilibrium profits of late entrants always exceed the expected profits of early entrants. In equilibrium the threshold quality of a firm declines in the threshold of the rival. The structure of the game is a *waiting game.*

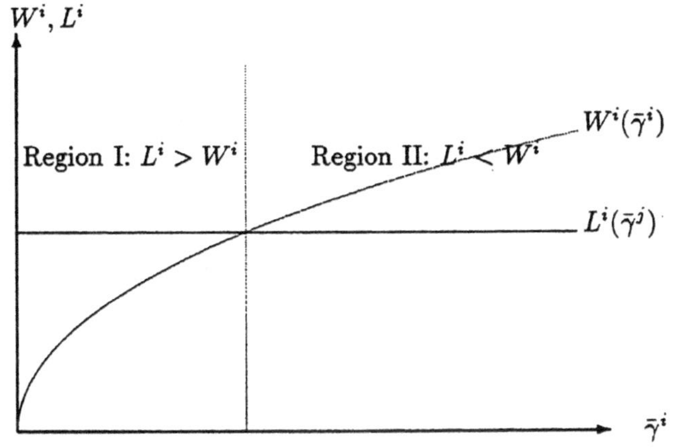

Figure 3.6: Different Regions of a Mixed Game

- The *third* class contains *mixed games.*[24] A mixed game arises if the difference between first-mover profits and second-mover profits changes its sign as the thresholds are varied. A qualitative example is depicted in Figure 3.6. Thus, a *mixed game* is defined as a game with *second-mover advantages* for low choices of first entry threshold levels, and with *first-mover advantages* for high first entry threshold levels.[25]

Fix $\bar{\gamma}^j$ at some level. Since $W^i(\bar{\gamma}^i)$ is increasing in $\bar{\gamma}^i$ by Assumption 3.4, while L^i is constant, W^i may cross L^i if it was below initially and if L^i is not too high. To the left of the crossing point the game has second-mover advantages, while it shows first-mover advantages to the right. Since L^i is decreasing in its argument, I can state the following theorem:

Theorem 3.3 *Take a mixed game of technological competition. (i) The borderline between first-mover advantages and second-mover advantages is a declining curve in*

[24]Dutta and Rustichini [1993] analyzed this class of games which they call "war of attrition cum pre-emption"-games (p. 750). They used a more restrictive setting in which there is an exogenous flow of technological knowledge into the industry, and is identical for all firms (a public good). In their analysis each firm has the same knowledge (and is therefore able to produce the same quality) at each point in time. See the discussion of Theorem 3.2 above.

[25]The two first classes can thus be interpreted as special cases of this third class.

the (γ^i, γ^j)-space for each firm. It is the geographical locus, where $W^i(\gamma^i) = L^i(\gamma^j)$ for $i = 1, 2$. This divides the (γ^i, γ^j)-space into three distinct regions.

(a) *In any equilibrium between the origin and the lower envelope of the borderlines for the two firms there are second-mover advantages. Reaction curves are decreasing.*

(b) *In any equilibrium above the upper envelope of the two firms there are first-mover advantages. Reaction curves are increasing.*

(c) *There are two subregions between the two borderlines. In each subregion there is one firm which prefers to be the first in the market, while the other prefers to be second.*

(ii) The reaction curves are continuous, if the profit functions in the market stage are all C^1. (iii) Equilibria may exist in any region. Assumptions 3.1–3.4 guarantee that at least one equilibrium exists.

Proof: (i) Note that the slope of the borderline for firm i is given by

$$\frac{d\gamma^i}{d\gamma^j} = \frac{(L^i)'(\gamma^j)}{(W^i)'(\gamma^i)}$$

which is negative by Assumption 3.4 and the previous discussion. The interpretation of the different regions is reflected in the slopes of the reaction curves in these regions. (ii) From the first order condition (3.19) one can conclude that there is no discontinuity if the thresholds are varied across the regions, unless the profit functions themselves have kinks. (iii) The first part is clear. Firm i chooses a positive threshold even if $\bar{\gamma}^j$ is zero because of Assumption 3.1. Because of Assumption 3.4 there is an implicit upper bound on optimal thresholds (because profits are bounded and research is costly). In addition, the game eventually is a game with strategic complementarities. Either there exists an equilibrium in this region, or there exists an an equilibrium in a lower region. For the details, see the proof of Theorem 3.1 in Appendix A.1.□

Interpretation: In Figure 3.7, four possible scenarios are depicted. For simplicity I assumed that the borderlines between the regions are linear and that firms are *ex*

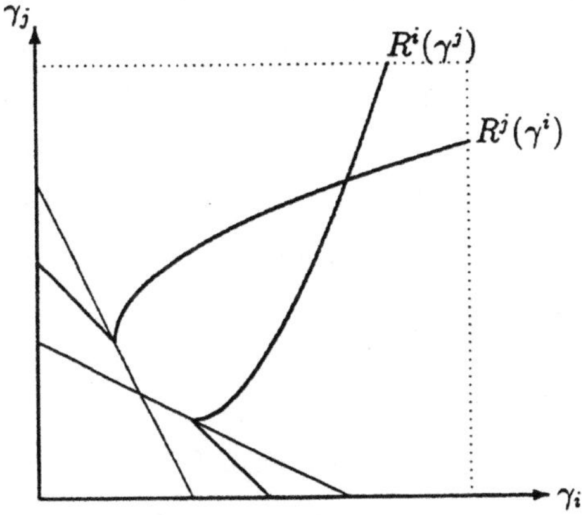

Figure 3.7 (a)

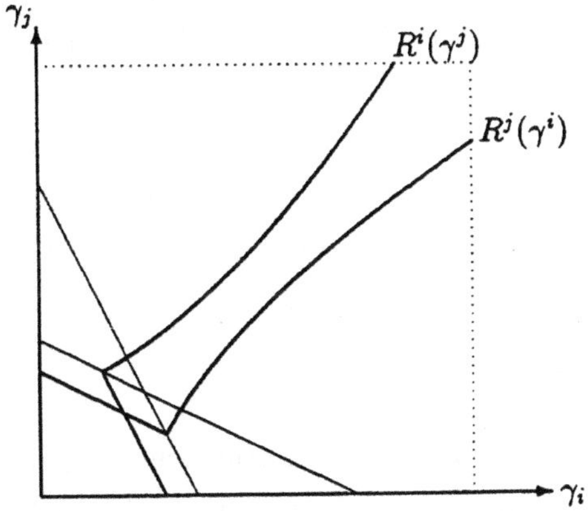

Figure 3.7 (b)

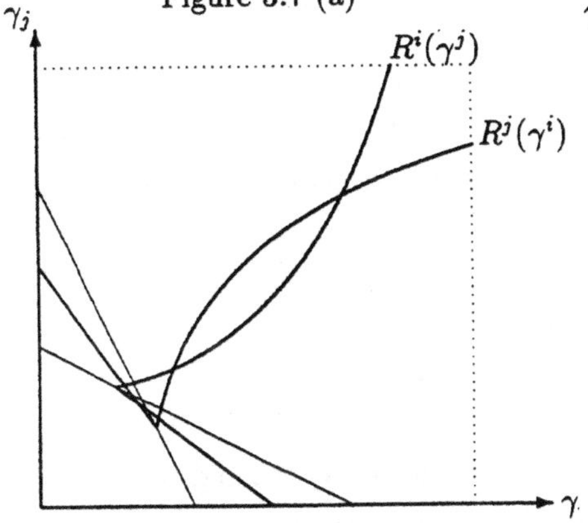

Figure 3.7 (c)

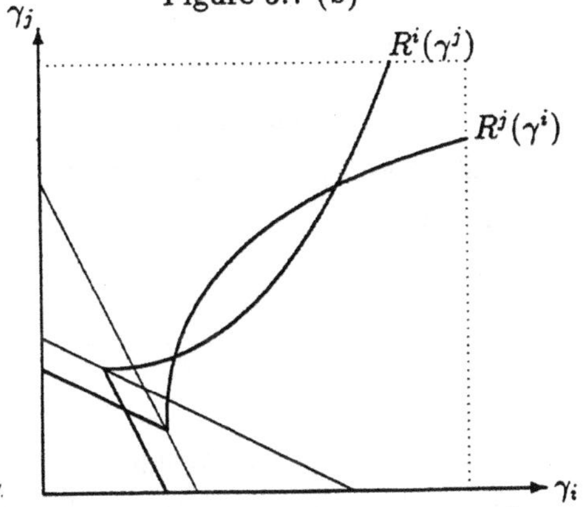

Figure 3.7 (d)

Figure 3.7: Possible Equilibrium Outcomes in "Mixed Games"

ante identical. The borderlines are represented by thin lines.[26]

In the first example (Figure 3.7 (a)) there is a unique equilibrium with first-mover advantages. Although the game eventually is a game with strategic complementarities, equilibria need not exist in the region with first-mover-advantages. Look at Figure 3.7 (b). In this example there is a unique equilibrium with second-mover advantages. However, equilibria may co-exist in every region, a case depicted in Figure 3.7 (c). In addition to the symmetric equilibria (one with first-mover, the other with second-mover advantages), two more asymmetric equilibria exist. One of the firms chooses a high threshold strategy in the first entry game, while the other firm chooses low threshold. The last example (Figure 3.7 (d)) shows a situation where only symmetric equilibria exist. There is a unique equilibrium with second-mover advantages as well as multiple equilibria with first-mover advantages.

The Value of Waiting

I now turn to the question which equilibria actually will be played. In a game with multiple equilibria it is natural to ask for a selection argument that reduces this indeterminacy. To discuss this point, let me concentrate on symmetric equilibria, i.e. like in Figure 3.7 (d).

Suppose there are two symmetric equilibria $\bar{\gamma}, \hat{\gamma}$, with $\bar{\gamma} < \hat{\gamma}$, and that the firms are identical. From equation (3.18) the expected profits from playing the equilibrium $\bar{\gamma}$ are given by

$$V(\hat{\gamma}, \hat{\gamma}) := \frac{-c + \lambda \int_{\bar{\gamma}} (W(\gamma) + L(\gamma)) dF(\gamma)}{r + 2\lambda(1 - F(\bar{\gamma}))} \tag{3.32}$$

What happens to the expected profits if firms move on to play the higher equilibrium $\hat{\gamma}$? Let the derivative along the 45°-line be denoted V_γ. One has

$$V_\gamma = \frac{2\lambda f(\gamma)(-c - r(W(\gamma) + L(\gamma)) + \int_\gamma (W(\zeta) + L(\zeta) - 2W(\gamma)) dF(\zeta)}{(r + 2\lambda(1 - F(\gamma)))^2}$$

[26] I have not been able to establish a result stating e.g. that the borderline of firm i always crosses the rival's borderline from above (or below). In Figure 3.7 the borderline of firm j is drawn steeper than the borderline for firm i. If the reverse would be the case, equilibrium outcomes like in Figure 3.7 (c) would not occur in a symmetric game. In this case, only strategically "pure" equilibria (with either first-mover advantages or second-mover advantages for both firms) may exist.

Using the first order condition (3.19) one obtains

$$V_\gamma = \frac{2\lambda f(\gamma)}{(r + 2\lambda(1 - F(\gamma)))^2}[W(\gamma) - L(\gamma)]r. \tag{3.33}$$

Thus I can state the following proposition:

Proposition 3.12 *Suppose that the firms are identical. Expected profits from technological competition are increasing along the 45°-line in the (γ^1, γ^2)-space in the case of first-mover advantages, while decreasing in the case of second-mover advantages.*

Remark: Proposition 3.12 stresses a general feature of games with strategic complements: In games with strategic complements profits increase if the players move to a higher equilibrium.

From the proof of Theorem 3.2 (see Appendix A.2) we already know that if $\bar{\gamma}$ is a symmetric equilibrium with second-mover advantages, another symmetric strategy combination $\hat{\gamma} > \bar{\gamma}$ can only be an equilibrium if the following condition is fulfilled:

$$\int_{\bar{\gamma}}^{\hat{\gamma}} (W(\hat{\gamma}) - L(\gamma))dF(\gamma) \geq 0. \tag{3.34}$$

Note that in the case where there are only equilibria with first-mover advantages this is no restriction. The condition (3.34) is necessary to guarantee that the expected profits from playing the higher equilibrium are indeed higher than the expected profits from playing the lower equilibrium. Thus (3.34) may rule out equilibria if $\bar{\gamma}$ is an equilibrium with second-mover advantages.[27]

To further restrict the set of possible equilibria I suggest the following reasoning that takes the dynamic nature of the game into account.

When firm 1 wants to play the equilibrium $\hat{\gamma} > \bar{\gamma}$ it somehow has to *coordinate* with firm 2. Suppose that they cannot communicate. In this case the firm may form

[27]The reason for this is that a mixed game with a symmetric equilibrium with second-mover advantages is no longer a simple *monotone stopping game*. In a *monotone stopping game* equilibria are in myopic strategies ("enter if the expected profit from entering now is higher than the expected profit from sampling exactly once more"). Simply looking for the first order conditions may therefore be misleading. One has also to compare the expected profits from different 'equilibria' as is reflected in (3.34). See Mamer [1987] for the monotone case.

beliefs on which equilibrium the rival will play. Let these beliefs be denoted p for the equilibrium $\bar{\gamma}$ and $(1-p)$ for $\hat{\gamma}$. Firm 1 therefore expects a payoff $pV(\hat{\gamma},\bar{\gamma}) + (1-p)V(\hat{\gamma},\hat{\gamma})$.

Now suppose that firm 1 maximizes against these beliefs. In addition, firm 1 is even more clever in that it takes into account that firm 2 entails the same reasoning. Such a firm enters directly whenever it realizes a quality γ, with $\bar{\gamma} \leq \gamma < \hat{\gamma}$ and $\gamma \geq \tilde{\gamma}$ where $\tilde{\gamma}$ is given by

$$W(\tilde{\gamma}) = pV(\hat{\gamma}, \max\{\bar{\gamma},\tilde{\gamma}\}) + (1-p)V(\hat{\gamma},\hat{\gamma}). \tag{3.35}$$

I write $V(\hat{\gamma}, \max\{\bar{\gamma},\tilde{\gamma}\})$ because also firm 2 has an incentive to coordinate away from the equilibrium $\bar{\gamma}$. It therefore prefers to wait until one of the firms realized at least $\hat{\gamma}$, as long as it does not lose expected profits given its beliefs. In this case firm 2 would enter if it realizes a $\gamma \geq \tilde{\gamma}$. At the same time firm 2 never enters with a quality inferior to $\bar{\gamma}$, since this is the lowest equilibrium. Therefore the maximum operator is introduced.

The probability that a firm realizes a quality γ with $\bar{\gamma} \leq \gamma < \hat{\gamma}$ and $\gamma \geq \tilde{\gamma}$ is given by

$$F(\hat{\gamma}) - \max\{F(\bar{\gamma}), F(\tilde{\gamma})\}, \tag{3.36}$$

and the conditional probability on that γ lies in the interval $[\bar{\gamma},\hat{\gamma}]$ is

$$\tilde{p} := \frac{F(\hat{\gamma}) - \max\{F(\bar{\gamma}), F(\tilde{\gamma}),\}}{F(\hat{\gamma}) - F(\bar{\gamma})}. \tag{3.37}$$

Note that $\tilde{p}$ is a good candidate for the beliefs the firms should form, since it is *consistent* with rational behavior: In an equilibrium where beliefs are formed according to this rule, a firm enters with probability $\tilde{p}$ with a quality from the interval $[\max\{\bar{\gamma},\tilde{\gamma}\},\hat{\gamma})$, and with probability $1-\tilde{p}$ with a quality $\gamma > \hat{\gamma}$.

It can be that $V(\hat{\gamma}, \max\{\bar{\gamma},\tilde{\gamma}\}) < V(\bar{\gamma},\bar{\gamma})$, since the firms deviate from equilibrium. At the same time it is clear that $V(\hat{\gamma}, \max\{\bar{\gamma},\tilde{\gamma}\}) < V(\hat{\gamma},\hat{\gamma})$, since $\hat{\gamma}$ is the equilibrium with the highest profits. Therefore it is possible that the $\tilde{\gamma}$ that solves equation (3.35) under the belief formation according to (3.37) is smaller than $\bar{\gamma}$. In this case coordination to the better equilibrium $\hat{\gamma}$ fails, and both firms play $\bar{\gamma}$. The equilibrium $\hat{\gamma}$ will in this case never be played. If $\tilde{\gamma} > \bar{\gamma}$ the firms coordinate to the higher equilibrium with positive probability.

Let me summarize the discussion:

Theorem 3.4 *Let there be a multiplicity of symmetric equilibria. Assume that the firms form beliefs about rival's strategies according to (3.37). (i) The lowest equilibrium is always played with positive probability. (ii) Consider two neighboring equilibria $\bar{\gamma}, \hat{\gamma}$ with $\hat{\gamma} > \bar{\gamma}$. The higher equilibrium is never played if $\tilde{\gamma}$ from equation (3.35) is smaller than $\hat{\gamma}$. (iii) If there is a symmetric equilibrium $\bar{\gamma}$ with second-mover advantages, then the condition (3.34) must hold for any other symmetric equilibrium.*

Remark: The beliefs formulated in this section are of course not the only beliefs firms can form. A firm can simply believe that the better equilibrium is always played. If both firms behave accordingly, no firm has an incentive to deviate. But let me argue that this form of belief is atemporal and does not take into account that firms have to wait if they want to play the higher equilibrium. In a dynamic sense belief formation according to (3.37) is more robust to the impatience of firms, and therefore more likely to shape real world behavior.

The Product Innovation Example of Dutta and Rustichini

To close this section, let me present the product innovation model in Dutta and Rustichini [1993]. This example obtains as a special case of the present model. They assumed that there is a unidimensional, increasing technology process for simplicity identified with time.

Firm i is the first mover (in equilibrium), while firm j is the second mover. Firms are identical and research costs are normalized to zero. The profit of the second mover does only depend on the difference between the entry times of the two firms. The second entry time is therefore always spaced away from first entry by the same amount of time, irrespective of the first entry time. Thus, second mover maximum profits are constant, and the profits of firm i do only depend on his entry decision. See Figure 3.8 for a graphical exemplification of their game.

Firm j always moves after t_m if firm i has not yet entered . Therefore firm i chooses its optimal entry time t^* from the interval $[0, t_m]$. and entry of the second firm

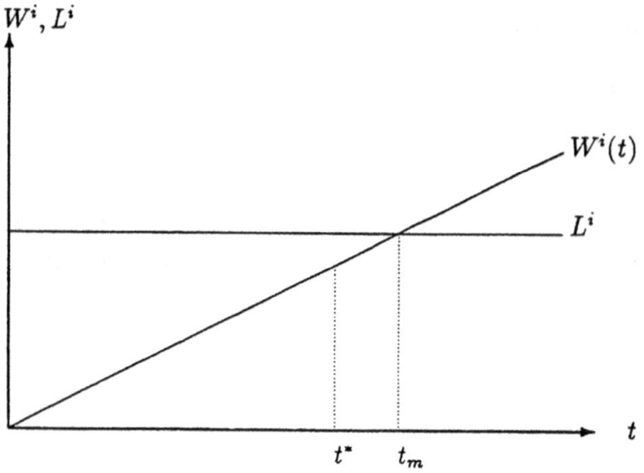

Figure 3.8: The Product Innovation Example of Dutta and Rustichini [1993]

is spaced away optimally. Dutta and Rustichini [1993] call this the *maturation equilibrium* "to emphasize the fact that a second mover waits optimally to innovate, makes higher profits by doing so and consequently does not dissipate the monopoly rents of the first mover" (p. 760).

3.5 Summary and Conclusion

In the present chapter I analyzed a model of dynamic technological competition between two firms. The technological standards (or quality choices) of firms are endogenously determined, as well as the order and time of entry into a market. Once in the market, firms are committed to the standard with which they entered the market.

Let me summarize the main results:

- The quality chosen by the firm that enters second is a function of the quality the first entrant has chosen. It is
  - (i) increasing in the quality of the first entrant if quality levels are *strategic complements* on the market stage.

(ii) decreasing in the quality of the first entrant if quality levels are ***strategic substitutes*** in the market stage and if the discount rate r is small.

- The *quality gap* decreases in the quality chosen by the first entrant, provided that the second entrant chooses a higher quality.

- The strategies chosen in the technological competition game for first entry are

  (i) *strategic complements* if and only if the equilibrium played entails *first-mover advantages.*

  (ii) *strategic substitutes* if and only if it entails *second-mover advantages*

- Equilibria with *first-mover advantages* are symmetric.

- Equilibria with *second-mover advantages* can either be symmetric or asymmetric. There is at most one symmetric equilibrium with *second-mover advantages.* This is the equilibrium that implies the lowest quality choices for the first entrant in the case that also other (symmetric) equilibria exist.

- *Spillovers* from early entrants to later entrants have ambiguous effects. They unambiguously increase second-mover profits. First-mover profits may both increase or decrease, dependent on the structure of the game in the market stage. The effect on the quality chosen by the first entrant is ambiguous, too.

- *Multiple symmetric equilibria* lead to coordination problems. Therefore entry tends to be early.

- If there is a symmetric equilibrium with *second-mover advantages*, this equilibrium is even more likely to be played. The reason is that in this case the monotone structure of the game can be destroyed, which may rule out other equilibria.

- If the firms form *consistent beliefs* to coordinate to higher (symmetric) equilibria with higher expected profits, the lowest equilibrium is always played with positive probability. It may also be that always the lowest equilibrium is played.

Comparison to the Literature

Although the model of Lippman and Mamer served as a starting point for the formal framework used here, the analysis and results go far beyond their findings. The possibility for post-innovation competition and the endogeneity of the firm's characteristics in the post-innovation stage allows for an analysis of the interdependence between post-innovation market structure and the incentives to innovate. I derived a close connection between a classification of the strategic structure of a game of technological competition and the relative expected profits in the market stage.

This differentiates the present study also from Dutta and Rustichini [1993]. They restricted the analysis to the case where new knowledge is a public good available for every firm in the industry, and with no idiosyncratic component. Under this assumption the game degenerates to a pure *game of timing.* The close connection between strategic structure and relative profits can no longer be analyzed.

Before I turn to the next chapter, let me emphasize two shortcomings of the present model. The *first* is that firms are committed to the quality with which they entered the market. This assumption should be relaxed (or at least derived as a result) in a next step.

The *second* is the close connection between the decisions to stop research and to bring an innovation to the market. This simplifies the analysis, since it restricts the strategy sets of firms to the choice of one single stopping time. In the next chapter, a simplified model with different informational assumptions is used to analyze the incentives of firms to space away the revelation of information away from the date the innovation is made.

A very interesting issue that is ignored here is the influence of financial constraints on the innovative choice of firms. In Chapter 5, a two-period model is developed to gain insights into this question.

Chapter 4

Strategic Use of Delay and Secrecy for Rent Protection in Technological Competition

4.1 Introduction

In the present chapter I analyze the incentives firms may have to postpone the commercial use of an innovation. This question naturally arises from the analysis in the last chapter, where I assumed that firms enter instantaneously after they finished their research project. I use here a simplified version of the stopping game introduced in the last chapter. The quality or technological standard of the innovation is given, but an additional source of uncertainty is introduced. In the setting discussed below, firms have incomplete information about the productiveness of the research technology that is used in the industry.

This idea of *hazard rate uncertainty* was first introduced by Choi [1991].[1] The consequence of this additional uncertainty is that firms get increasingly pessimistic about

[1] D. Soskice pointed out to me that the uncertainty introduced in this paper may also cause strategic delay in oil drilling. Suppose two firms have claims in a region where an oilfield is expected. A firm that is successful in finding oil may have incentives to hold back the information about the success, because drilling is expensive and often drilling firms have to give up because they run out of funds. If the rival is not successful, holding back the information about own success may drive the rival sufficiently pessimistic to drop out.

the prospects of further investment into the R&D-project and therefore sometimes abandon research before the task is completed. Thus, a similar pattern as in the investment model of Kamien and Schwartz [1971] (see also the reformulation by Grossman and Shapiro [1986]) arises, but without the assumption of a finite investment horizon.

Motivation

Let me formulate *two stylized facts* that led to the present analysis.

- Many research projects are abandoned before completion.
- Innovation shelving occurs.

Discussion: The first stylized fact is inconsistent with standard optimal investment theory. As long as there is no new outside information, rational firms do not abandon an investment project in an infinite horizon model whenever they found it *ex ante* profitable to start at all.

An exception to this literature is the variable investment model of Kamien and Schwartz [1971]. The task firms try to solve is of unknown complexity, and is only rewarded if completed under a finite investment horizon. This has as a consequence that the continuation value of investment is not constant over time. Especially, the continuation value may decrease. In this case there is a point in time, after which also the optimal investment policy is decreasing. Eventually, it may fall down to zero before the task is completed.

The same effect can be generated if not uncertainty about the complexity or the difficulty of the task, but uncertainty about the productivity of the research strategy pursued is introduced into the model. In this case firms cease investment in finite time even in the case of an infinite horizon.

Although not explicitly taken into account below, financial restrictions are likely to reinforce the problem of cease-investment before task completion (I come back to this problem in the next chapter). Let me provide some examples.

- **The Superconducting Super-Collider (SSC)** When the CERN accelerator at Geneva started to work in July 1989, this was a large step ahead in high energy physics. To conduct research on the microstructure of material highly accelerated electrons are needed as "weapons" to provoke the decay of a atomic nucleus that serves as a target. The performance of particle accelerators is measured in electron volt (*eV*). The "storage ring" technology used by CERN made it possible to break through the barrier of a trillion *eV*.

Physicists try to check the correctness of explanations of a common law behind the three basic forces of attraction. Although high energy physics achieved a very high level, the financiers (governments) became increasingly pessimistic about the prospects of further investments in to even stronger accelerators. This led to the cancellation of the *SSC*-project (Dallas, Fort Worth) which would have allowed for a performance in the range of 20 trillion *eV*.[2] Inspite of that SSC was partly constructed by then, the authorities decided to cease financial aid in the early nineties.

- **Plutonium Based Reactor Technology** German authorities spend a couple of billions on the construction of a *plutonium based reactor* in Kalkar. It was conjectured that this technology would allow for a higher degree of effectiveness in nuclear energy production. Under the public pressure in the eighties and nineties, the german authorities became more and more pessimistic about the storage possibilities for the high toxical plutonium that is produced in the process of energy generation. A few years ago, the authorities decided to cancel the whole project of plutonium based reactor technology in Germany.

- **Superconductivity** In 1986, the later Nobel laureates Müller and Bednorz discovered a ceramic substance that was superconducting at 35° Kelvin. Superconductivity is an important research fields in physics, material sciences and process engineering because of the high resistance of common materials used for energy conduction. Superconductivity is a phenomenon observed in certain materials at extreme temperatures. It is the emergence of sluices on the atomic level where electrons can be channeled through almost resistant-free.

  Research expenditures slowed down over time because the financiers became quite pessimistic about the possibility of producing superconducting material at temperatures that allow for profitable commercial use. The findings of Bednorz and Müller "gave a whole new meaning to superconductivity" and "sparked a feverish race among scientists worldwide to find other compounds

[2]See Ayres [1994] for a history of particle acceleration.

that are superconducting at even higher temperatures"(Choi [1991], p.597). One might conjecture that without the findings of Bednorz and Müller investment into this technology had approached a zero level after some time.

- **Nuclear Fusion** To domesticate the power of the sun is a dream that motivated many researchers (and governmental or supragovernmental authorities) to spend money and effort on the controlled fusion of hydrogen nuclei to produce helium and energy. It was conjectured that the successful fusion would solve all energy problems of modern societies. Over time successes in this field have been quite modest, because fusion can only be conducted in a plasma environment with an temperature around 100 million degrees Celsius.[3] Fusion can only be controlled using gigantic electromagnets producing an electromagnetic field that contains the super hot plasma. Although the potential benefits from successful fusion are incredible, the financiers became reluctant to spend money on this type of research. And it is certainly not purely by chance that british researchers announced the successful conduct of fusion (for a small fraction for a second – then the electromagnetic field collapsed) short before E.E.C. authorities planned to run down the budget for this kind of research a few years ago.

The second stylized fact is that innovations are sometimes shelved. Again, let me provide at least one example.

- **Superconductivity** Let me return to the story of Bednorz and Müller. What did they do with their findings? – Choi [1991], p. 608 reports that they did not publicly announce their success, and instead submitted "a modest scientific paper to an obscure german journal [...] in order to continue their research in peace." One may conjecture that they believed to succeed to generate superconductivity at a higher temperature themselves, and therefore tried to avoid to attract competitors into the field.

Incentives to Hold Back Information

Let me turn to the connection between the two stylized facts. If the fact that firms or research institutions (or their financiers) become pessimistic on the prospects of

[3] At least with the known technologies. All attempts to conduct *"cold fusion"* turned out to be complete failures.

their own research projects happens to be true, we are in a situation where "to know that someone has solved a problem is to know a great deal" (Dasgupta and Maskin [1987]).

Traditional patent race analysis (see the Sections 2 and 3 in Chapter 2) suggests that the information of rival success is bad information for firms. In multi-stage races this may lead to dropout behavior. Bag and Dasgupta [1995] show in a model that is close to the analysis below that this result can be reversed if the information structure is endogenous and *hazard rate uncertainty* is introduced.

The fact of innovation shelving is substantiated in the theoretical literature by the *replacement effect* (See Chapter 2.3 for a discussion). I argue that the generation of *informational spillovers* may be a second source for innovation shelving.

Two sorts of informational spillovers are possible. *First*, the information generated by the fact that "someone has solved a problem" may be valuable for rival firms if there exists considerable uncertainty about the productivity of research projects, and if there is sufficient similarity between the approaches of different firms in an industry.

The *second effect* is that technological knowledge embodied in the innovation leaks out to rivals if the innovation is made publicly known. I already discussed some implications of technological spillovers for technological competition in the last chapter. Spence [1984], or Cohen and Levinthal [1989] presented theoretical treatments of the implications of technological spillovers on the incentives to innovate. The literature so far concentrated on the role of spillovers for the formation of research joint ventures.[4] The strategic role of those spillovers has been analyzed by de Fraja [1993]. Firms may deliberately choose to disclose technological knowledge to rivals if the innovations are complements in the product market. If innovations are substitutes in the product market the question of foreclosure of information becomes relevant.

In the next sections I concentrate on the possible implications of the first sort of informational spillovers. The remainder of this chapter is organized as follows. I give a short review on the literature on strategic information revelation in the next section. In Chapter 4.3 the model with uncertainty about the research technology is discussed. Two interrelated decision problems are treated, the timing decision to

[4]See e.g. d'Aspremont and Jaquemin [1988].

bring a completed innovation to the market if patent protection is incomplete, and the decision to abandon an unfinished research project if there had been no observed success for a longer period of time. Chapter 4.3 contains the comparative static analysis and a short (informal) discussion of the welfare consequences of strategic delay. Chapter 4.5 summarizes and concludes.

4.2 Revelation of Information and Delay

There is a growing literature in a number of fields dealing with the optimal behavior of agents or firms when the true state of nature is unknown initially but is learned over time.

Perhaps the most prominent examples are the studies about the adoption of *innovations of unknown profitability.* These include Jensen [1982], Spatt and Sterbenz [1985] or Bhattacharya, Chatterjee and Samuelson [1986].

In the models by Jensen [1982] and Spatt and Sterbenz [1985] information about the profitability of innovation has the form of publicly observed signals which come in successively over time. Thus, firms learn over time the true profitability of the innovation. Jensen [1982] concentrates on the diffusion profile of the innovation when firms differ in their priors about the profitability of the innovation, while Spatt and Sterbenz [1985] stress the strategic adoption decision under a "winner takes all"-assumption.

In the latter model, the basic tradeoff which shapes the adoption decision is that on one hand firms get better informed about the true profitability of the innovation if they wait longer, but on the other hand they risk to be preempted by less patient rivals. In the chosen framework there exist multiple equilibria which can be pareto-ranked from the earliest to the latest equilibrium,[5] and the set of equilibria shrinks to the earliest one if the number of firms goes to infinity. Similar to the case of multiple equilibria with a *first mover advantage* discussed in the previous chapter, the outcome is a *coordination game.*

Closer to the model I am going to present below is the approach taken by Bhattacharya, Chatterjee and Samuelson [1986]. They assume that firms do not differ in

[5] Early adoption is less profitable than later adoption in expected terms. Firms would therefore prefer to coordinate to an equilibrium where adoption occurs late.

their priors about the profitability of the innovation, but that they receive private signals about the true profitability of the innovation. Firms therefore may have an incentive to wait with the adoption decision in order to learn from early users about the true profitability. Two effects arise in their model. First, there is a bandwagon effect that leads to clusters of adoptions if one firm decides to adopt the innovation. Second, there is the effect that a firm tends to wait longer whenever its rivals do not adopt. The reason for this behavior is a *second-mover advantage* coming from the informational spillovers created by early adopters. Since private signals are not reliable, firms prefer that others reveal their information about the profitability of the innovation before they themselves adopt.

Mariotti [1992] provides an analysis of the decision to adopt an innovation by identical firms. The innovation can either be good (positive expected profits), or bad (negative expected profits). The true quality is revealed only after at least one firm adopted the innovation. With identical firms the equilibrium involves mixed entry strategies such that there is a positive probability at each point in time that the innovation is not adopted by any firm, and therefore a positive probability that the innovation is never adopted even if the total expected profit from adoption is positive.

A similar approach is taken in the *investment model* by Chamley and Gale [1994]. As in Bhattacharya, Chatterjee and Samuelson [1986], firms (or agents) differ in their information about the true state of nature. Chamley and Gale give an example of an economy that is in recession. Firms hesitate to invest because they have to rely on private information, e.g. about future demand conditions. They may want to wait until they observe other firms to invest even though they received favorable signals, since this reveals that these other firms received good signals, too. Again, this may either lead to a clustering of investment decisions or to inefficient delay of investment by all firms due to insufficient information.

Delay of the adoption of innovation or of investment decisions is also an important phenomenon in the presence of network effects and competing standards, as is pointed out e.g. by Farell [1987] and Farrell and Saloner [1988]. Early adoption of a standard may be favorable because of first mover advantages, but there are countervailing forces because there is the possibility that a different standard eventually becomes dominant. Coordination is favorable in such situations. Bolton and Farell [1990] analyze a situation of natural monopoly and two possible entrants which have private information about their production costs. Each firm that enters the market has to pay a non-recoverable entry cost, but only the low cost firm can survive. Given private information about own cost, firms tend to inefficiently postpone their entry decision compared to the full information case if they both have relatively high production costs. If both firms have relatively low production costs there is inefficient duplication.

A common feature of all studies discussed above is that *firms may find it profitable to wait* with their innovation adoption decisions. The incentives to postpone the entry or investment decision have two sources: The first source is the informational externality provided by the entry decision of other agents or rivals which renders it profitable to move second. The second source is an exogenous flow of information which mitigates the risk from taking the wrong decision as time goes by.

In my model there is a similar positive informational externality from *innovative entry* as in the models by Bhattacharya, Chatterjee and Samuelson [1986], Mariotti [1992] or Chamley and Gale [1994]. The entry decision of a firm reveals to rivals that the state of nature is favorable. Without such an information rivals tend to become increasingly pessimistic about future prospects and eventually decide to irreversibly quit competition. Therefore this time a firm has incentives to foreclose its information over a certain time interval to protect its rents from the innovation.

In a recent article, Bag and Dasgupta [1995] pursued a similar idea. They analyzed a two-stage race in the spirit of Lippman and McCardle [1987], and show that with the technological uncertainty described above, and with additional private information about the type of the firms it may be profitable to hold back the information about first stage success. This contrasts the findings of Lippman and McCardle, where intermediate success is likely to cause dropout of the rival. In the original version of Lippman and McCardle therefore no incentives exist to postpone the announcement of intermediate success, while it may be profitable if additional uncertainty is introduced.

4.3 Strategic Delay if the Research Technology is Random

In the last chapter I introduced a stopping game in which firms choose technological standards (or product qualities). The standards chosen together with the possibility of sequential entry determined the profits of the firms in equilibrium. The choice of technology and the entry time of a firm was closely connected: I assumed that a firm enters the market directly after finishing the R&D-stage. This connection is broken up here. To highlight the tradeoff that leads to deferral of entry in equilibrium a simplified version of the game in the last chapter is used here.

Setup: I analyze a model with two firms. The decision to defer entry does not primarily depend on whether they are identical or not. Thus, for most of the analysis they are assumed to be identical. The firms spend on a R&D-project at a constant rate c. There is a setup cost k. This cost has to be incurred before research is started. It is also due when a firm re-enters after a break. The projects of the two firms are identical, but it is assumed that the realizations are independent from each other. The firms have incomplete information about the true nature of the project. The research approach used in the industry can either be productive or not. In the first case firms successfully complete the project with instantaneous probability $\lambda^h dt$, while in the second case the probability is zero.[6] The assumption that both firms use the same research technology is restrictive. It can be relaxed at the price of additional complexity to correlated research technologies in the sense that the information about rival's success is of positive informational value for a firm. This positive value is generated whenever a firm can learn about its own success probability from the observation of rival's success.

Similar to the last chapter there is uncertainty concerning the *duration* and concerning the *outcome* of the project. The difference is that in the previous chapter the second source of uncertainty was the *quality* of the innovation the firms obtain, while in this chapter the possible outcomes degenerate to *success* or *failure*. There is no perfect patent protection. Thus, I allow for subsequent entry.

Information: Two polar cases are considered. *First*, firms can observe the success of rivals (or can conclude from equilibrium behavior if the rival has succeeded).

[6]This is the idea of *hazard rate uncertainty* introduced by Choi [1991].

This case is called the *perfect-monitoring case*. *Second*, firms cannot observe rival's success. This is the *no monitoring case*. The information structure of the game will be determined endogenously. This is the main innovation of the present chapter.

Strategies: Firms choose entry times as strategies. They invest into R&D, and conditional on success they decide when to enter the market. The entry decision of the firms will determine the informational structure of the game. This is so because I assume that the information about rival's success is signaled through the commercial use of the innovation.[7]

Profits: A monopolist earns flow profits rM. With two firms in the market duopoly flow profits are rD. The discount rate r is strictly positive. To hold notation simple, I assume that both firms earn identical profits in the market competition stage, irrespectively of the order of entry.

4.3.1 The Investment Decision of a Single Firm

I begin the analysis with the case of a monopolistic firm. The prior probability the firm attaches to having the productive technology λ^h is given by $p := p(0)$ $(= Prob\{\lambda = \lambda^h | t = 0\})$. Over time it updates this probability according to Bayes' rule:

$$p(t) = Prob\{\lambda = \lambda^h | \text{no success until time } t\} = \frac{pe^{-\lambda^h t}}{pe^{-\lambda^h t} + 1 - p}. \tag{4.1}$$

The instantaneous subjective success probability is given by $\lambda(t)dt = p(t)\lambda^h dt$. I assume that it is profitable for the firm to start the research project:

Assumption 4.1 *A firm finds it profitable to start research at* $t = 0$, *i.e. the following condition is fulfilled:*

$$c \leq \lambda(0)M - k$$

[7] Two alternative interpretations are possible: *First*, the decision to apply for patent protection: The application signals to potential rivals that a potentially profitable innovation is possible, and therefore may attract competitors. This has to be traded off against the value of legal protection. *Second*, the announcement of intermediate success in a multi-stage research process. See Bag and Dasgupta [1995] for details on this interpretation. They show that the endogenization of the informational structure in the multi-stage game of Lippman and McCardle (see Chapter 3.3) can change the results considerably.

As a preliminary result I state

Lemma 4.1 *The subjective posterior probability a firm attaches to having the productive research technology conditional on no success is a strictly decreasing function of time.*

Proof: Differentiating equation (4.1) with respect to t gives

$$\frac{\lambda^h p(p-1)e^{-\lambda^h t}}{(pe^{-\lambda^h t}+1-p)^2} < 0.\square$$

The time of completion of the research project is exponentially distributed with parameter $\lambda(t)$ (seen from time t). Let the date of completion be denoted τ. The expected expenditures of the firm from time t until completion are given by

$$\begin{aligned} -cE[\int_0^\tau e^{-rt}dt] &= -c\int_0^\infty \int_0^\tau e^{-rt}dtdF(\tau) \\ &= -\frac{c}{r}\int_0^\infty (1-e^{-r\tau})dF(\tau) \\ &= -\frac{c}{r}E[1-e^{-r\tau}] \\ &= -\frac{c}{r}(1-\frac{\lambda(t)}{r+\lambda(t)}) \\ &= \frac{-c}{r+\lambda(t)}, \end{aligned} \tag{4.2}$$

where $E_\tau[1-e^{-r\tau}] = \frac{\lambda(t)}{r+\lambda(t)}$ is the appropriate discount factor for the to calculate the present value for the future monopoly profit streams.[8] Suppose that the firm had no success until time t. It continues research as long as expected profits from doing so are positive. The value from continuation of research beyond time t is given by

$$V = \frac{-c+\lambda(t)M}{r+\lambda(t)} \tag{4.3}$$

Thus, the monopolistic firm continues research as long as

$$c \leq \lambda(t)M. \tag{4.4}$$

Therefore I can state

[8] See Chapter 2.2.2 for a derivation.

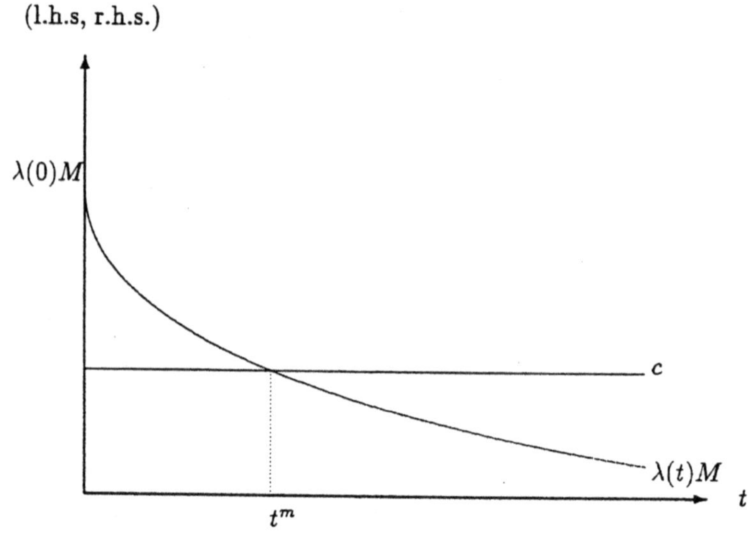

Figure 4.1: Determination of the Exit Time for a Single Firm

Proposition 4.1 *The monopolistic firm continues on the research project as long as inequality (4.4) holds. There is a unique time t^m for which (4.4) holds with equality. Conditional on not having been successful until t^m the firm abandons research.*

The result is depicted in Figure 4.1. Over time the firm gets more and more pessimistic about the prospects of future success over time, and therefore eventually decides to abandon its project if no success has yet been observed.

4.3.2 Information Revelation and Technological Competition

Let there be two firms which are engaged in technological competition. They may differ in their efficiency λ_i^h, $i = 1, 2$. Firms undertake research project that are similar, i.e. if one firm learns that its rival has made a discovery it may conclude that its own research technology is productive, too. The first firm that completed the innovation and brings it to the market earns monopoly profits rM. After both firms entered the market, duopoly profits rates are rD. In the present setting I always assume $D < M$.

The firms share common priors $p = p(0)$ about having the productive technology. The posterior probabilities are formed according to Bayes' rule. Note that in the case of two firms the subjective posteriors of a firm depend on whether it can monitor its rival or not. In the *perfect monitoring* case the posteriors are given by

$$\begin{aligned} p(t) &= Prob\{\lambda_i = \lambda_i^h | \text{no success until time } t,\ i = 1,2\} \\ &= \frac{pe^{-(\lambda_i^h+\lambda_j^h)t}}{pe^{-(\lambda_i^h+\lambda_j^h)t} + 1 - p}. \end{aligned} \tag{4.5}$$

In the case where *the rival cannot be monitored* the posteriors are formed according to

$$\begin{aligned} p_i(t) &= Prob\{\lambda_i = \lambda_i^h | \text{no own success until time } t\} \\ &= \frac{pe^{-\lambda_i^h t}}{pe^{-\lambda_i^h t} + 1 - p}. \end{aligned} \tag{4.6}$$

In the sequel, the subscript for the firm is suppressed wherever not essential for the analysis. I write $p(t)$ for the posteriors if not explicitly stated otherwise. This simplification of notation should not lead to confusion, since it is only relevant in a later stage of the analysis, when the time period is derived over which firms defer the commercial use of an innovation. Note that the following is true:

Lemma 4.2 *Firms get increasingly pessimistic over time in case of no success. In the perfect monitoring case the subjective posterior probabilities firms attach to having the productive technology decrease faster as in the case without monitoring.*

Proof: Differentiate equations (4.5) and (4.6) with respect to time.□

No Strategic Delay

There are two cases to consider: *First*, the case in which a firm that completes an innovation directly makes commercial use of it (short: *direct entry-case*) and *second*, the case where firms decide to defer entry to a later date. I consider the *no strategic delay*-case first.

The first firm that enters the market with its innovation earns instantaneous monopoly flow profits rM until the second firm enters, and both firms earn duopoly flow profits rD thereafter. The expected profits of firm i conditional on being first are

$$\begin{aligned} W^i &:= \int_0^\infty \lambda_j^h e^{-\lambda_j^h t} \left[\int_0^t rMe^{-rs}ds + \int_t^\infty rDe^{-rs}ds\right] dt \\ &= \frac{rM + \lambda_j^h D}{r + \lambda_j^h}, \end{aligned} \tag{4.7}$$

while the expected profits of firm i conditional on being second are given by

$$L^i := \int_0^\infty \lambda_i^h e^{-\lambda_i^h t} \left[\int_0^t (-c) e^{-rs} ds + \int_t^\infty r D e^{-rs} ds\right] dt = \frac{-c + \lambda_i^h D}{r + \lambda_i^h}. \tag{4.8}$$

Let $\tau = \tau_i \wedge \tau_j$ be the time of first innovation. τ is exponentially distributed with parameter $\lambda_i + \lambda_j$. The appropriate discount factor is thus given by

$$E_\tau[e^{-r\tau}] = \frac{\sum_{i=1,2} \lambda_i(t)}{r + \sum_{i=1,2} \lambda_i(t)}. \tag{4.9}$$

The conditional expectation given that i preempts its rival is given by

$$E_{\tau_i}[e^{-r\tau_i} | \tau_i < \tau_j] = \frac{\lambda_i(t)}{r + \sum_{i=1,2} \lambda_i(t)}. \tag{4.10}$$

Total expected profits are therefore

$$V^i(t) := \frac{-c + \lambda_i(t) W^i + \lambda_j(t) L^i}{r + \lambda_i(t) + \lambda_j(t)} \tag{4.11}$$

Let me assume that the firms always find it profitable to start R&D at $t = 0$ and, once a firm abandoned its research project it will find re-entry always unprofitable. In addition, let it be always profitable for the firm to continue research if it can be sure that it has the productive research technology. Thus, I impose the following assumptions:

Assumption 4.2 *All firms find it profitable to start research in $t = 0$, i.e.*

$$c + k \leq p(\lambda_i^h W^i + \lambda_j^h L^i).$$

Assumption 4.3 *Re-entry into the research project after a break is unprofitable, i.e.*

$$c + k > \lambda_i^h D \quad , i = 1, 2.$$

Note that a firm may only want to re-enter research if the information about the research technology changes due to rival success. Assumption 4.3 therefore states that duopoly profits are too low to attract re-entry.

Assumption 4.4 *Continuation on the project is always productive if the research technology is productive with probability one, i.e.*

$$c < \lambda_i^h D \quad ,i = 1,2.$$

Firms continue research unless $V^i(t)$ is negative for one firm. This firm drops out, while the other firm stays in as long as $c < \lambda_i(t)M$. In addition to the previous assumptions let me impose an indifference rule:[9]

Assumption 4.5 *A firm stops investing into the project whenever it is indifferent between continuation of the research project and exit.*

If the firms differ in their efficiency parameters λ_i^h the exit order is determined by relative efficiency. The dates upon which the firms abandon the research project if they do not succeed are deterministic. If the firms are identical, two types of equilibria exist.[10] There are two asymmetric equilibria with one firm exiting, and an equilibrium in mixed strategies in which firms randomize over the decision to stay in or to exit the research project.[11]

Strategic Delay

I turn now to the second case in which firms may *strategically foreclose information.* Think about the following decision problem of a firm which successfully completed an innovation. It can either directly introduce its innovation into the market[12] or it can wait until either the rival is pessimistic enough to abandon his project or succeeded, too. Let me assume in a first step that the firms always find it profitable to wait with the commercial use of the innovation until some time $\bar{t}$. This point in time $\bar{t}$ is chosen such that both firms are indifferent between continuation and exit conditional on no observed success. I conduct the analysis for a given $\bar{t}$. In the next step (see the section on the role of posterior formation) $\bar{t}$ is endogenously determined. I restrict the analysis to the case with identical firms.

[9]This assumption is equivalent to restriction to pure entry strategies. Otherwise an "open-set problem" arises that implies a continuum of mixed strategy-equilibria.

[10]See Tirole [1988], p. 311 on the *war of attrition*.

[11]This is exactly the effect in symmetric waiting games used by Mariotti [1992] to show that there may be unused innovations under competition even if there are positive expected profits.

[12]Or, apply for an (imperfect) patent in the alternative interpretation of the model.

Assume that firm i completes its innovation project at time t. Although it loses monopoly profits it decides to wait with the commercial use of the innovation until $\bar{t}$[13], which is the exit time of the rival in case of no success. The rival firm j may or may not complete its project in the meantime. In the present setting this is not observable by firm i, since its rival also forecloses the information and waits in case of success.

The probability of rival success in the time interval $[0, \bar{t}]$ is $exp(-\lambda^h \bar{t})$. If the rival succeeded, they both earn a flow profit rD after $\bar{t}$. If not, firm i earns monopoly flow profits rM. Thus, the expected profits of firm i conditional on success in time $t < \bar{t}$ are

$$\begin{aligned} \pi(t, \bar{t}) &:= e^{-r(\bar{t}-t)} \left[D \int_0^{\bar{t}} \lambda^h e^{-\lambda^h s} ds + M \int_{\bar{t}}^{\infty} \lambda^h e^{-\lambda^h s} ds \right] \\ &= e^{-r(\bar{t}-t)} [D + (M - D) e^{-\lambda^h \bar{t}}]. \end{aligned} \tag{4.12}$$

Interpretation: The discount factor reflects the interest losses from waiting. Because firm i knows that it has succeeded in t, it also knows that the other firm has the productive technology. Thus, the instantaneous success probability that i assigns to firm j is $\lambda^h dt$. The first integral reflects therefore the probability that firm j successfully completes the project in the time interval $[0, \bar{t})$, too. In this case, the firms share the market after $\bar{t}$. The second integral reflects the probability that firm j fails during $[0, \bar{t})$, and therefore exits. In this case firm i earns monopoly profits after $\bar{t}$. The conditional profits given that firm i succeeded in t and defers until $\bar{t}$ do not depend on $p(t)$, since it knows for sure that the productive technology is used in the industry.

The following lemma proves to be useful in the equilibrium analysis:

Lemma 4.3 *The expected profits conditional on own success $\pi(t, \bar{t})$ in the case with strategic delay are increasing in t.*

Proof: Obvious.

[13] I use a fixed $\bar{t}$ here. Of course, the time of exit of an unsuccessful firm depends on the formation of its posteriors, and therefore on the informational structure. This informational structure will later be endogenously determined from the equilibrium behavior of firms. See Theorem 4.1 and Proposition 4.3 below.

Unconditional expected profits (at time t) from undertaking research are therefore given by

$$V(t,\bar{t}) := \frac{-c + \lambda(t)\pi(t,\bar{t})}{r + \lambda(t)}. \tag{4.13}$$

Interpretation: Before firm i succeeds, it does not know that it has the productive technology. Therefore it uses

$$\frac{\lambda(t)}{r + \lambda(t)}$$

as the appropriate discount factor for future expected profits. If it succeeded, it earns conditional expected profits $\pi(t,\bar{t})$.

4.3.3 Equilibrium Analysis

Suppose that firm i completes its innovation at some time t. It has to decide whether to bring its innovation to the market without delay or to wait until the other firm either abandons its research project or enters with a completed innovation itself.

Let the exit time conditional on no observed success for firm j be fixed at some $\bar{t}$. I will later determine this time endogenously from the equilibrium conditions. If firm i enters directly, its expected profits are given by W^i in equation (4.7). In the case of identical firms this is

$$W := \frac{rM + \lambda^h D}{r + \lambda^h}. \tag{4.14}$$

From Lemma 4.3 expected profits conditional on success in the case of *strategic delay of entry* are increasing in the time of success. Suppose that initially they are lower than expected profits from *direct entry* after success. Since expected profits from direct entry are constant over time, there may exist a point in time t^* at which a firm is indifferent between direct entry and delay of entry. Since the firms are identical, also t^* is identical for both firms. Let me assume for the moment that $t^* < \bar{t}$. The firms would then enter directly before t^* conditional on success before t^*, while they defer entry to $\bar{t}$ whenever $t^* < t \leq \bar{t}$. Equation (4.12) has therefore to be slightly changed, since delay occurs only after t^* (and not always, as was assumed above). It becomes

$$\pi(t, t^*, \bar{t}) := e^{-r(\bar{t}-t)}[D + (M - D)e^{-\lambda^h(\bar{t}-t^*)}]. \tag{4.15}$$

Let me now show that $t^* < \bar{t}$. Look at the limit case, where t^* approaches $\bar{t}$ from below:

$$\lim_{t^* \uparrow \bar{t}} \pi(t, t^*, \bar{t}) = Me^{-r(\bar{t}-t)}.$$

But since $t^* < t \leq \bar{t}$ in the case of strategic delay, also t has to approach $\bar{t}$. Thus,

$$\lim_{t^* \uparrow \bar{t}} \pi(t, t^*, \bar{t}) = M. \tag{4.16}$$

Since it always holds that $M > W$, it follows that expected profits from delay are eventually higher than from direct entry. Thus, we have $t^* < \bar{t}$. If the profits from delay of entry are higher at $t = 0$, it follows that $t^* = 0$ and equation (4.12) and (4.15) coincide.

A firm chooses to directly enter the market whenever

$$\frac{rM + \lambda^h D}{r + \lambda^h} \geq e^{-r(\bar{t}-t)}[D + (M - D)e^{-\lambda^h(\bar{t}-t^*)}]. \tag{4.17}$$

From the previous discussion I can state

Proposition 4.2 *Fix the exit time for firm $j(i)$ at $\bar{t}$. Let the two firms be identical, i.e. if firm $i(j)$ finds it profitable to wait with introduction of an innovation, the same is also true for firm $j(i)$. There exists an unique time $t^* < \bar{t}$ at which firm $i(j)$ is indifferent between instantaneous commercial use of an innovation completed at that time and waiting until $\bar{t}$. It enters directly with an innovation completed before t^*, and defers entry after t^*.*

Proof: Directly from equation (4.17). Before t^* the expected profit from direct entry is higher, while deferring leads to higher expected profits after t^*, given that also firm j decides to wait after t^*.□

From the symmetry assumption it can be concluded directly that both firms defer entry after t^* if an exit time $\bar{t}$ exists. Until now I assumed that there is such a time.

The next step is therefore to show that the firms drop out in equilibrium after $\bar{t}$ if they did not observe a success in the time span before $\bar{t}$. Fix the behavior of firm i as before.

Firm j had no success until $\bar{t}$. In $\bar{t}$ it has to decide whether to exit or to continue investing. Suppose that firm j continues a little bit longer until $\bar{t}+dt$. It spends an amount cdt and observes with probability $1-exp(-\lambda^h(\bar{t}-t^*))$ that firm i succeeded and enters the market with its innovation. With probability $exp(-\lambda^h(\bar{t}-t^*))$ firm i drops out and firm j could establish a monopoly position in case of success. It completes the innovation project with instantaneous probability $\lambda(\bar{t})dt$.

Thus the expected profits from continuing research after $\bar{t}$ are exactly $V(\bar{t},\bar{t})$, which are zero by the definition of $\bar{t}$. Firm j therefore is indifferent between dropping the research project and continuation. By Assumption 4.5 it drops out.[14] Thus, $\bar{t}$ exists. As a consequence the following theorem is stated for the symmetric case:

Theorem 4.1 *For a given $\bar{t}$ there exists a unique equilibrium with strategic information foreclosure. If one of the firms succeeds early ($t < t^*$), it directly brings its innovation to the market. The other firm continues researching until success. If one of the firms, or both, complete the innovation during $[t^*,\bar{t}]$ the commercial use is deferred until $\bar{t}$. If no firm succeeds, both drop out at $\bar{t}$.*

The Role of Posterior Formation

In this section the exit time $\bar{t}$ is endogenously determined. To do this I return to the two cases of posterior formation in the *perfect monitoring case* and the *no monitoring case*. The date $\bar{t}$ at which firms drop out conditional on no observed success is given through equations (4.12) and (4.13) by

$$c = \lambda(\bar{t})[D + (M - D)e^{-\lambda^h(\bar{t}-t^*)}]. \tag{4.18}$$

The determination of $\bar{t}$ according to equation (4.18) is depicted in Figure 4.2. Note that the time $\bar{t}$ is dependent on $\lambda(\bar{t}) = \lambda^h p(\bar{t})$. To close the model (through the endogenous determination of the time $\bar{t}$) I need to know how the firms determine their posterior probabilities about having the productive technology.

[14] In an earlier version of this model I assumed that only one firm acts strategically. In this case the profit function of the non-strategic firm has a discontinuity at $\bar{t}$, which leads to equilibria in mixed strategies. I thank Jay Pil Choi for this remark. However, no discontinuity arises if both firms act strategically.

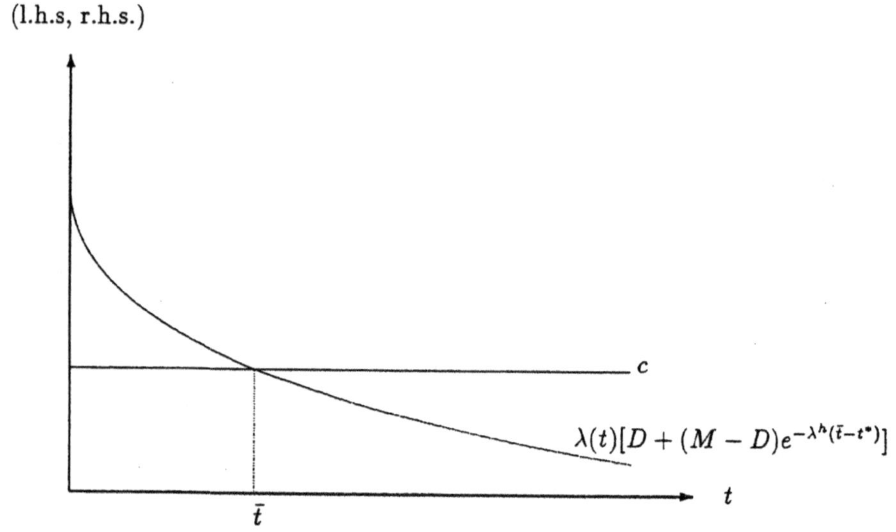

Figure 4.2: Exit Time with Strategic Foreclosure

In the case with *perfect monitoring* firms choose the updating rule (4.5) to form posteriors, while they update according to equation (4.6) if there is *no monitoring.* I assume that firms use *consistent* belief formation in equilibrium, i.e. they include all available information. Therefore they may infer that the rival has not succeeded until t^* if no market entry occurred until t^*. However, the behavior of firms after t^* is not informative, since no entry occurs during the interval $[t^*, \bar{t}]$, irrespectively of success by either firm. Let me summarize the discussion in the next proposition:

Proposition 4.3 *Firms may conclude in equilibrium that the rival has not yet succeeded before t^* if the rival has not introduced its innovation, while the behavior of the rival is not informative after t^*. In an equilibrium with consistent belief formation, the posteriors $p(t)$ are formed according to (4.5) if $t < t^*$, while they update according to (4.6) thereafter.*

4.4 Comparative Statics

In the present model firms decide whether to defer the commercial use of an innovation conditional on success, and how long they pursue a research project if not successful. The main result is that unsuccessful firms exit at a fixed date $\bar{t}$, and that all innovations in a time interval $[t^*, \bar{t}]$ are shelved and introduced after $\bar{t}$.

I turn now to the comparative static analysis on how the exit date and the delay period depend on the parameters of the model.

Length of the Delay Period

Remember that the time t^* after which firms choose to wait with the introduction of an innovation into the market is given by

$$\frac{rM + \lambda^h D}{r + \lambda^h} = e^{-r(\bar{t}-t^*)}[D + (M - D)e^{-\lambda^h(\bar{t}-t^*)}]. \tag{4.19}$$

Note from equation (4.19) that the length of the delay period $(\bar{t} - t^*)$ does not depend on the absolute value of $\bar{t}$. This simplifies the comparative static analysis since changes in the optimal exit date $\bar{t}$ according to changes in the parameter set do not influence the length of the delay period. The following theorem summarizes the comparative static results concerning the delay period. The results are depicted in Figure 4.3.

Theorem 4.2 *The delay period $(\bar{t} - t^*)$ is shortened by an increase in the discount rate r as well as in the monopoly profits M. An increase in duopoly profits D as well as in the efficiency parameter λ^h leads to a longer delay period.*

Proof: See Appendix A.4.

Remark: It is clear from equation (4.19) that the length of the delay period does not depend on the flow cost of research c.

Discussion: The results in Theorem 4.2 are quite intuitive. An increase in the discount rate r increases the loss from the delay of the commercial use of an innovation. If monopoly profits increase firms are less willing to give up the transitory monopoly opportunity to protect future profits. This effect is mitigated if duopoly profits are

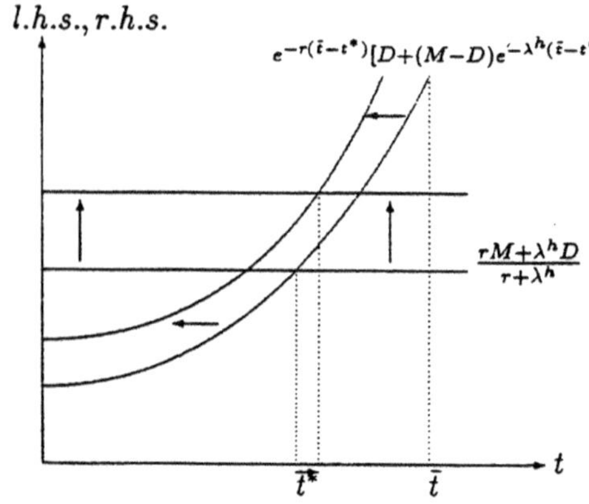

Figure 4.3 (a): Increase in M

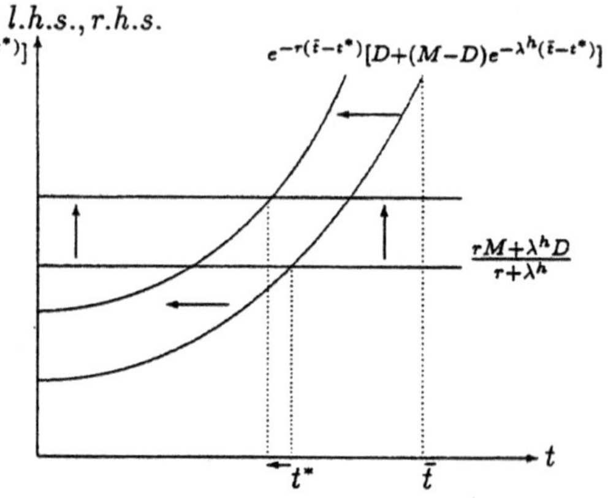

Figure 4.3 (b): Increase in D

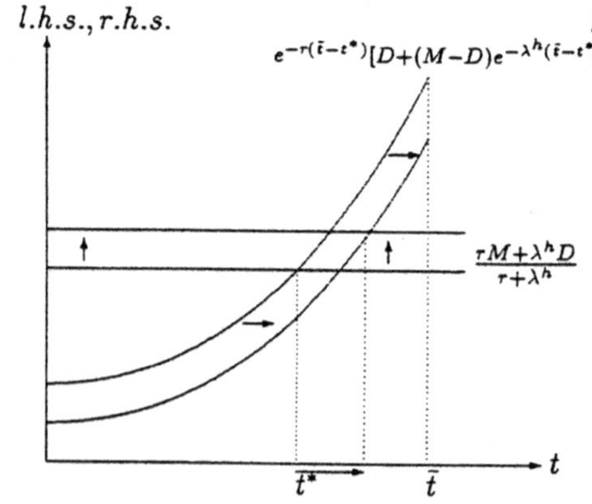

Figure 4.3 (c): Increase in r

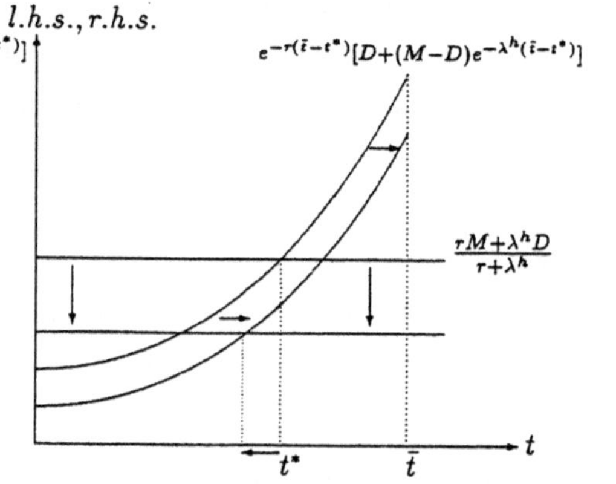

Figure 4.3 (d): Increase in λ^h

Figure 4.3: Comparative Statics of the Delay Period $\bar{t} - t^*$

higher. The effect of the efficiency parameter λ^h is ambiguous and depends on its relative size compared to the discount rate r.

The effects of relative profit changes are depicted in Figure 4.3 (a) and (b). Note that there are two countervailing effects at work, since an increase in both monopoly and duopoly profits leads to an increase in total expected profits with and without strategic delay. Higher expected profits from direct introduction *ceteris paribus* reduce the optimal delay period, while higher expected profits from deferment work in the opposite direction. Nevertheless it can be concluded from equation (4.19) that the overall effect is negative for a change in monopoly profits, while positive for a change in duopoly profits.

A change in r has a clear-cut effect, as can be seen in Figure 4.3 (c). A higher discount rate increases profits from direct entry and reduces the expected profits from deferment. Thus, both effects work towards a shorter delay period.

The effects of a change in λ^h are depicted in Figure 4.3 (d). Since the relevant profits in equation (4.19) are expected profits conditional on own success, the efficiency parameter λ^h in (4.19) is the instantaneous success probability of the rival. It therefore measures the *intensity* of technological competition.[15] A higher rival efficiency unambiguously reduces expected profits, no matter whether from direct or later entry with the net effect being positive. Thus, more intense technological competition leads to a lengthened delay period.

The Exit Time

Let me now turn to the comparative statics concerning the exit time $\bar{t}$. Remember that the exit date is determined by

$$c = \lambda(\bar{t})[D + (M - D)e^{-\lambda^h(\bar{t}-t^*)}]. \tag{4.20}$$

There are two effects to consider. *First*, the direct effect of a change in the set of parameters on the exit date $\bar{t}$ from equation (4.20). *Second*, there is an indirect effect coming from the change in the delay period $\bar{t} - t^*$ due to parameter changes. Note that equation (4.20) does not depend on the discount rate r. From a change

[15]A higher efficiency parameter λ^h leads to a higher success probability of the rival. In a generalization to n identical firms rival success would occur with instantaneous probability $(n-1)\lambda^h$ (and is therefore increasing in n). This case is not treated here.

in r there is consequently only an indirect effect on $\bar{t}$. The results are summarized in Theorem 4.3 and illustrated in Figure 4.4.

Theorem 4.3 *The exit date $\bar{t}$ is increasing in monopoly profits M as well as in the discount rate r. It declines in the research expenditure rate c. The total effect of a change in the efficiency parameter λ^h is indeterminate in the symmetric model: An increase in own efficiency has a positive effect on bart, while the effect from an increase in rival's efficiency has the opposite effect. Also a change in duopoly profits D leads to an indeterminate effect. An increase in the delay period leads to earlier exit.*

Proof: See Appendix A.5.

Discussion: A higher expected monopoly profit naturally extends the time span during which the firms have incentives to continue on their research projects, while a more costly project is abandoned earlier.

The efficiency parameter λ^h enters equation (4.20) in two ways: *First*, through $\lambda(\bar{t}) = \lambda^h p(\bar{t})$. This is the posterior the firm attaches to the instantaneous probability of own success. An increase in λ^h therefore increases the probability of own success. At the same time, the success probability of the rival enters directly into equation (4.20). An increase in rival's instantaneous success probability leads to a decrease in expected profits since the rival has a greater chance to complete its innovation as well.

The discount rate does not directly influence the exit date of a firm, but there is an indirect effect coming from a shortened delay period.

The effect of a change in duopoly profits is not so easy to understand. The direct effect comes from an increase in expected total profits. The indirect effect works in the opposite direction, since the delay period is increasing with higher duopoly profits. The negative effect of a lengthened delay period on the exit date stems from the increased subjective probability attached to rival success during the delay period which tends to decrease expected profits in case of own success.

The effects are illustrated in Figure 4.4. The direct effects of parameter changes stem from the changes in expected profits from continuation on the research project. The indirect effects come from changes in the delay period analyzed in Theorem

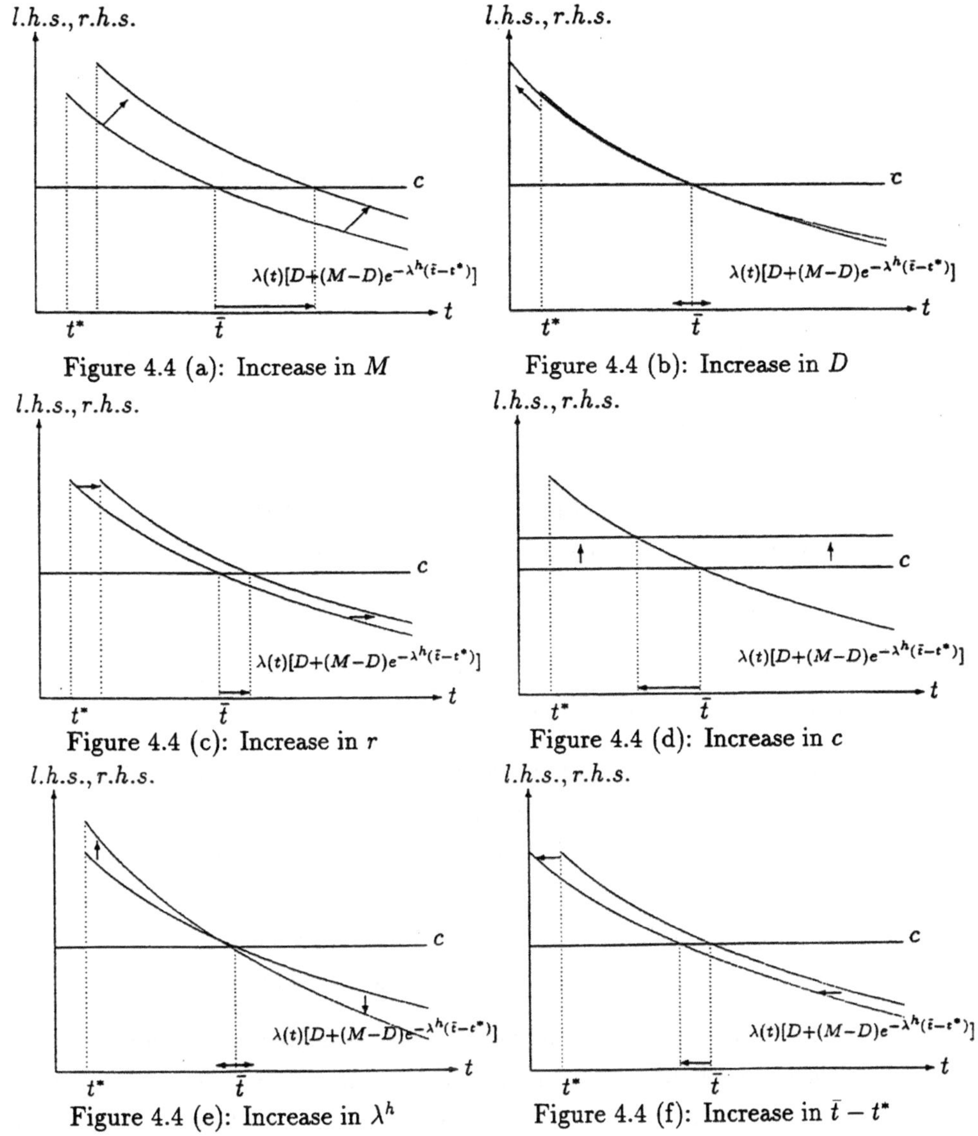

Figure 4.4 (a): Increase in M

Figure 4.4 (b): Increase in D

Figure 4.4 (c): Increase in r

Figure 4.4 (d): Increase in c

Figure 4.4 (e): Increase in λ^h

Figure 4.4 (f): Increase in $\bar{t} - t^*$

Figure 4.4: Comparative Statics of the Exit Time $\bar{t}$

4.2. The direct effects are illustrated by vertical shifts of the expected profits from continuation (or research costs). The indirect effects are visualized through shifts in t^* for fixed $\bar{t}$, and thus represented by horizontal shifts in Figure 4.4.

As in the previous case, the effects of profit changes are illustrated in Figure 4.4 (a) and (b). Both effects work into the same direction for a shift in monopoly profits. The total effect is therefore positive. As stated, the total effect from a change in duopoly profits is indeterminate. Only the indirect effect is working if the interest rate is changing (Figure 4.4 (c)), while a change in research costs c (Figure 4.4 (d)) has only a direct effect which leads to earlier exit. For changes in λ^h the direct and and also the total effect is indeterminate. I therefore left out the indirect effect in Figure 4.4 (e) and concentrated on the direct effect. An increase in λ^h leads to higher expected profits if $\bar{t}$ is low, and to lower expected profits for higher $\bar{t}$. The exact point around which the curve is twisted may be both to the right or to the left of the actual exit date $\bar{t}$. Therefore no definite conclusions can be drawn. To understand how the indirect effect is working, I illustrated the *pure* indirect effect of a change in the length of the delay period in Figure 4.4 (f).

Is Strategic Delay Beneficial?

Let me conclude the analysis of the present chapter with a short discussion on the benefits and losses from strategic delay.

Strategic delay leads to the effect that if both firms are unsuccessful, they invest longer into R&D than in the case where this opportunity does not exist. The reason for this effect is that rivals do not learn from each other about the prospects of further investment into the research project during the delay period. They tend therefore to be overly optimistic concerning their chances to successfully complete the project. From the viewpoint of society this is a waste of resources.

A second effect is that a successful innovation is not marketed directly for *predatory* reasons. The firms hold back information about own success and refrain from market entry to discourage rivals from further investment. Whenever only one firm succeeds during the delay period this leads to post-innovation monopoly. This has two welfare effects: *First*, consumers have to wait before the innovation is marketed. *Second*, direct entry by the successful firm signals to the rival that the research technology is productive. The rival would then not drop out, and eventually enters the market, too. Both effects, the delay effect and the effect that decreases competition in the post-innovation market are detrimental to consumer welfare.

On the side of the firms, strategic delay is clearly advantageous in expected terms. This advantage comes from the increased probability of a monopoly position in the post-innovation market. Taking into account that monopoly does imply a deadweight loss for total welfare, this private gain turns into a social loss when consumer surplus is considered, too.

The overall effect of strategic delay is therefore likely to be detrimental to welfare. Thus the question arises, under which circumstances strategic delay is less likely to occur.

An answer is provided by the previous comparative static analysis: Stronger rent protection for innovators (*i.e.* through better patent protection) diminishes the incentives to undertake predatory actions. But this is only one side of the medal. Stronger patent protection again leads to a prolonged monopoly situation in the market. Therefore one should carefully think about other (non-market) incentives for firms to make their results publicly known. The self-interest of researchers is a good starting point. Lead time may sometimes be valued higher by the scientific community than it is in the marketplace.

4.5 Summary and Conclusion

In the present chapter I provided an analysis of the incentives to foreclose information about successful innovation (by delay of market entry) in a model of technological competition between two firms. The driving force behind the result is a source of uncertainty coming from incomplete information about the productivity of the research strategy pursued by firms.

Let me summarize the main results:

- Uncertainty about the productivity of the research strategy used by a firm leads to the effect that the firm gets increasingly pessimistic about the prospects of further investment. Eventually the firm therefore stops investing conditional on no success.

This result complements the theoretical results from optimal investment theory. It can explain a slowdown (or cease) in investment without external provision of information or the assumption of a finite investment horizon.

- Firms have an incentive to shelve innovations in the presence of uncertainty about the research process whenever re-entry costs of rivals are sufficiently high. The time period over which innovations are shelved increases in the intensity of competition.

- Strategic delay makes unsuccessful firms race longer. This comes from the fact that they are overly optimistic about the prospects of further investment because they cannot observe that rivals are unsuccessful, too. Thus, there is a waste of resources because of insufficient information flows between firms.

- The overall effect of strategic delay is likely to be detrimental to social welfare. The decision to defer innovative entry is taken for purely *predatory* reasons.

Before I turn to the next chapter I want to point out some shortcomings of the present approach, as well as some points of departure for future research.

The analysis undertaken here has a very normative character. If there is uncertainty about the productivity of research and if research undertaken by different firms or institutions is similar in nature, there is an incentive to strategically foreclose information. It does not claim that firms always use this opportunity or that it is always an opportunity at all. Empirical evidence is coarse, and therefore one should be careful in the interpretation of the results.

However, there is recently a growing literature on delay of investment in both micro- and macroeconomics in the case where firms differ in information, or if firms have incomplete information about the true state of nature in the future, but learn over time (See Section 2 of the present chapter). All work into this direction ignores (to my knowledge) competition effects on the timing of investment (or adoption of innovation or market entry). Markets are either competitively organized or there is post-adoption (or post-entry) monopoly. The model in the present chapter points to the possible implications of competition between firms in the presence of informational spillovers.

Finally, I said much about the implications of outside finance for the incentives to take predatory actions, but did not incorporate this topic into the formal analysis. Some of the consequences of financial constraints are discussed in the next chapter.

Chapter 5

Innovation under Financial Constraints and Competition

5.1 Introduction

In this chapter I analyze a simple two-period model of innovation with a financially constrained firm that is dependent on outside funding. It therefore writes a contract with an outside investor. The basic structure of the model is similar to Bolton and Scharfstein [1990]. I assume the same informational structure between the contracting parties (*i.e.* profits are not verifiable to courts), but depart from their analysis in two decisive directions.

First, the firms choose to exert effort to complete a given research project. Therefore the incentives to innovate can be analyzed. Second, I provide a richer framework for the market competition outcome. In Bolton and Scharfstein profits are either high or low. In my model profits depend on the success of innovative activity of both firms. Thus, there are more (four) states of nature stemming from the possible combinations of firms' successes in each period (both firms succeed, only one firm succeeds, no firm succeeds). This extension seems to be innocuous, but it changes the analysis considerably.[1]

[1]The main impact of this extension on the analysis is that public announcement of the contract cannot serve as a commitment device against a rival firm when renegotiation is possible. This was possible in the Bolton and Scharfstein model. See the analysis below for details.

Motivation

Most firms finance their R&D-projects by retained earnings. To raise outside funds is often difficult for two reasons:

- *First*, R&D-projects are characterized by technological and entrepreneurial risks which are dependent on the ability of researchers and their effort. This creates moral hazard and adverse selection problems for potential lenders, especially when firms are protected by limited liability.
- *Second*, potential lenders normally are not able to correctly assess the profit opportunities firms have after successful completion of a project. In addition, because of legal institutions and limited obligation to report relevant internal data to the public, lenders in many cases have to rely on reported figures as a contractual basis. To give a firm incentives to report relevant data truthfully is costly, and therefore may lead a reduction in the firm's liquidity.

These constraints to raise outside funds work as a disadvantage especially for small and innovative firms. Nelson [1959] pointed out that large and diversified firms may have comparative advantages to innovate because they are able to internally finance their projects and to insure against individual failures by diversifying the risk of individual projects.

Empirical Evidence

Empirical studies confirm this reasoning. Himmelberg and Peterson [1994] used the Spence [1979] model of firm growth and find strong evidence that young and small firms are rationed in the capital market. One important reason is that the only collateral those firms have to offer is the value of their growth opportunities, which is not assessable and verifiable by outside claimholders.

In another study, Holtz-Eakin *e.a.* [1994] draw similar conclusions: The survival probability of a firm increases significantly when liquidity constraints are relaxed. Opler and Titman [1994] report that leveraged firms which have a high R&D intensity (as well as small firms and firms in highly concentrated industries) are struck harder in times of industry downturn than are their less leveraged and less research intense rivals. Evidence in the same direction is provided by Phillips [1995]. He states that "leveraged firms lose market share or fail to gain share when small rivals exit the industry." It seems to be that leveraged firms "compete less aggressively".

Financial Constraints and Predation

I argue that there is not only a comparative advantage for established firms to innovate but that they also may have incentives to use their position to prey young and innovative rivals out of the market by giving them a hard time to find long-term finance.

This is consistent with the findings of Opler and Titman [1994]. Leveraged firms in concentrated industries experience relatively high losses in industry recession. Opler and Titman conjecture that "concentration also proxies the gains associated with removing a weakened competitor [...]" (p. 1029).

Kovenock and Phillips [1993, 1995] report a related result within an empirical study of American industries in which the leverage of some firms changed considerably in the past. Firms that increase their leverage by LBO's or MBO's tend to reduce total investment and close plants, while their competitors tend to do the opposite.

The importance of financial structure is now widely acknowledged in studies of managerial strategy, competition and organizational design. Among the first that pointed to the importance of financial structure to constrain management behavior were Jensen and Meckling [1976] and Myers [1977]. Building on their work, Brander and Lewis [1986] published an article which is now a standard reference in the Industrial Organization literature when the impact of financial structure on firm behavior in oligopoly situations is discussed.

The main result of their paper is that *leveraged* firms, protected by *limited liability*, choose to behave *more aggressively* in output markets.[2] This is not in accordance with the effect observed by Kovenock and Phillips.

Bolton and Scharfstein [1990] posed the problem in a slightly different form. While Brander and Lewis stressed the point that firms might use leverage as a commitment device in competition (by constraining the possible set of actions for the manager), Bolton and Scharfstein discussed the possibility that leveraged firms are more vulnerable to rival attacks, if refinancing of a leveraged firm is conditioned on reported success, and rivals can inflict first period losses on the leveraged firm in a multiperiod framework. A similar idea is pursued here.

The remainder of the chapter is organized as follows. In the next section a review

[2]For a similar model that analyzes optimal contracts see Fershtman and Judd [1987].

on related studies is given. Especially the basic structure of situation in which standard debt is optimal is discussed. In Section 5.3 a simple two-period model of innovation and market competition is presented. It is a simplified (repeated) version of the Dasgupta and Stiglitz [1980b] model with random success and only two firms. The optimal contract and the consequences for the behavior of firms is analyzed in Section 5.4. Two informational structures are distinguished. The contract can either be unobservable but rationally inferable, or observable. When the contract is unobservable, predation occurs. The observable contract may be a no-predation contract, but it is likely to be not renogotiation proof. Section 5.5 concludes.

5.2 FINANCIAL CONSTRAINTS IN OPTIMAL CONTRACTS

A number of studies analyzed the optimal structure of financial contracts between borrowers and lenders under asymmetry of information. Townsend [1979] considers a model with costly state verification for the lender.

Although Townsends' model is a consumption exchange model, let me present the idea in an investment framework. There are two points in time: At $t = 0$ a debt contract is signed. The borrower needs outside funding to use an investment opportunity that pays back randomly at $t = 1$. The project is *ex ante* profitable, and both parties share the same information at the point in time the contract is signed. The realization of the project is privately observed by the borrower, but can be verified by the lender if she incurs a verification cost. The optimal contract in this framework specifies a verification region: The borrower announces at $t = 1$ her realized profit to the lender. If the profit is relatively high (higher than some threshold value) the lender receives this threshold value (which corresponds to the upper bound of the verification region). If the reported profit is below the lender incurs the verification cost and is paid back the full realized profit.

The shape of this contract resembles a standard debt contract. There is constant repayment in case of high profits. If the profit is lower than the agreed repayment, the firm is liquidated and the liquidation value goes to the lender.

A similar situation is analyzed in Gale and Hellwig [1985]. If the lender chooses to verify the state (profits are announced in the verification region) the firm is liquidated. This leads to a non-pecuniary cost (bankruptcy cost) imposed on the firm. The lender incurs a verification cost that is both dependent on the state and on the amount invested. Gale and Hellwig solve for the optimal incentive compatible contract in this situation under the assumption of a competitively organized banking sector. The result is again a standard debt contract.

Diamond [1984] has no verification technology in the basic version of his model, but if there is a shortfall from agreed repayment the lender is able to incur a non-pecuniary loss on the borrower.[3]

In all models presented so far the optimal contract is a standard debt contract. This seems to be the standard result in the considered type of agency problems. Firms have private information about the realization of some relevant variable and cannot *ex ante* commit themselves to truthfully reveal this value. This allows the firm to extract the surplus above some threshold if profit realizations are high, but at the cost of punishment or liquidation in case of low profits.

There are remedies to the financial restrictions the firms face. Bester [1987] introduces the idea of collateral to alleviate the rationing problem. In Diamond [1991] reputational issues are discussed. Monitoring is another device to alleviate agency problems, but is often costly.[4] Another possibility is to allow for control rights of lenders within the firm. This idea is put forward e.g. in Aghion and Bolton [1992] or Dewatripont and Tirole [1994].

Beside these remedies the standard structure of the contract involves a *stick* and a *carrot*. The stick is either the liquidation of the firm or some non-pecuniary penalty. The carrot comes from the fact that in an incentive-compatible contract the firm is able to extract some surplus if profits are high. In Bolton and Scharfstein [1990] (and also in the model treated below) the *stick* is provided in a multiperiod framework by the threat of discontinuation of finance.

[3]Note that the enforcement of such a contract is unwarranted since renegotiation is possible. Diamond does not consider this issue. Hart and Moore [1989] look at the renegotiation problem in a multiperiod framework. Again the optimal contract involves standard debt. See also Hart and Moore [1994].

[4]For a recent study on the role of monitoring and debt capacity see Holmström and Tirole [1993].

5.3 A Two-Period Model of Innovation and Competition

5.3.1 The Model

In the present chapter I analyze a simple model of technological competition with financial constraints. One firm is dependent on outside funding, while the other is able to finance research projects out of retained earnings. Outside funding is provided by rational investors (banks). The two firms only differ in their ability to internally finance the projects and are otherwise identical. Call the *leveraged firm* firm i, while the other *unleveraged rival* is called firm j.

They are competing in a product market. The model has two time periods, period 1 and period 2. In each period there is an innovation opportunity for each firm. Operating on the market and using the innovation opportunity costs an amount F in each period. This fixed cost F must be paid before the project is started. Firm i has to write a contract with an investor to raise F. The investor writes an optimal incentive contract with the firm that is funded. In doing so, she takes into account that firm i is protected by limited liability. The firms and the investor are assumed to be risk-neutral.

The innovation project leads to a success with probabilities θ_i, and θ_j for the two firms. The success probabilities depend on the effort the firms spend on the project. The effort is represented by a private cost (disutility) function $g(\theta_i)$, $i = 1, 2$. This private cost is assumed to be increasing ($g'(\theta_i) > 0$) and convex ($g''(\theta_i) > 0$) in the success probability.

Information: The investment game in each period is a simultaneous-move game. Both firms know that firm i is dependent on outside funding. The terms of the contract are private information of firm i. However, firm j may rationally infer that an optimal incentive contract is written between firm i and the investor. I will later relax this assumption to analyze the consequences of publicly announced contracts.

The investor knows the basic data of the market, i.e. he knows the *ex ante* possible payoff realizations of the firm it finances. In addition, I assume that the investor knows the private cost function $g(\theta_i)$. However, the realizations of the profits as well as the actual effort spent by the firms are not verifiable by the investor (and therefore not contractible). I show below that even with this detailed knowledge of the industry's data the investor provides insufficient finance. Introducing additional asymmetry of information (*e.g.* about the private cost function or about the possible states of nature) would presumably worsen this problem.

Strategies: (i) The investor makes a *take it or leave it*-offer to firm i prior to investment. Thus, I assume that the bargaining power over the contract is with the investor. The offer contains an incentive contract that is based on reported profits. (ii) The firms choose their effort levels simultaneously at the beginning of each period. It is assumed that the two projects are independent, *i.e.* there are no spillovers between firms. (iii) At the end of the first period firm i reports its realized profits to the investor. (iv) The provision of second period finance is then based on these reports.

Profits: Profits are serially uncorrelated over the two periods.[5] The profits of firm i are given by the following matrix:

$i \backslash j$	success	no success
success	π^i_{11}	π^i_{10}
no success	π^i_{01}	π^i_{00}

I concentrate on the problem of firm i. The profits of firm j are symmetric (*i.e.* $\pi^i_{k,l} := \pi^j_{l,k}$, $k, l = 0, 1$). Superscripts are suppressed for notational convenience.

The following assumptions are assumed to hold throughout the analysis:

Assumption 5.1 *The following relation between profits holds for firm i:*

$$\pi_{10} > \pi_{11} > \pi_{00} > \pi_{01}.$$

The assumption that the profits are higher if both firms succeed as in the case where none has a success is consistent with Cournot-competition. Assume linear inverse

[5] I also analyzed the case of serially correlated profits in an earlier version. Little additional insights were gained, at the price of heavily increased notational complexity. The basic structure of the optimal contract derived below carries over to the case with serially correlated profits. However, the investor will no longer enforce optimal effort when profits are serially correlated. This makes the computation of the exact terms of the contract more difficult. The decisive feature of the contract, namely that there is not always finance provided in the second period if reported profits are low, remains valid.

demand functions, i.e. $p = A - x$ and constant marginal costs c. For the symmetric case, duopoly profits are $\frac{(A-c)^2}{9}$, thus negatively dependent on c. If the innovation is cost-reducing and competition is Cournot with linear demand and constant marginal costs, profits are higher in the case where both firms have successfully completed an innovation.

Assumption 5.2 *With positive probability firm i cannot pay back its loan,* i.e. $F > \pi_{01}$.

Without this assumption there is no serious finance problem. It is therefore a natural assumption if one is interested in the consequences of financial constraints.

Let the expected first period equilibrium profit of firm i be denoted $\bar{\pi}$.

Assumption 5.3 *For the expected first period profit $\bar{\pi}$ it holds that $\pi_{11} > \bar{\pi} > \pi_{00}$.*

This assumption is a restriction on the effort levels chosen in equilibrium.[6] It can be relaxed without seriously changing the analysis. It turns out that the optimal repayment scheme specifies that whenever reported profits are higher than expected profits, the investor is paid back $\bar{\pi}$, while he gets the whole profit whenever a profit below $\bar{\pi}$ is reported. Assumption 5.3 is then a convenient means towards a simple characterization of the optimal contract.[7]

5.3.2 Equilibrium Analysis without Financial Constraints

Firm i chooses its effort to maximize net profits in both periods. Since profits are assumed to be serially uncorrelated, each period can be treated separately and the optimal effort is the same in both periods. It solves

$$\max_{\theta_i} \quad \theta_i\theta_j\pi_{11} + \theta_i(1-\theta_j)\pi_{10} + (1-\theta_i)\theta_j\pi_{01} + (1-\theta_i)(1-\theta_j)\pi_{00} - g(\theta_i) - F. \tag{5.1}$$

[6] One can easily show that it is equivalent to impose a lower bound on the equilibrium effort levels, where this bound is given by $g'(\theta) > \pi_{00} - \pi_{01}$ in a symmetric equilibrium.

[7] In a generalization of the present model one could assume a continuum of possible realizations of profits from some interval. An equivalent assumption to Assumption 5.3 would then be that interior realizations of profits occur in equilibrium with positive probability. See *e.g.* Faure-Grimaud [1995].

The expression to be maximized is the expected first period profit of Firm i net of effort costs g and the fixed operating cost F. The first order condition for this problem is

$$g'(\theta_i) = \theta_j(\pi_{11} - \pi_{01}) + (1 - \theta_j)(\pi_{10} - \pi_{00}). \tag{5.2}$$

Since $g'(\theta_i) > 0$ and $g''(\theta_i) > 0$, an optimal solution exists. Firms choose positive levels of θ in equilibrium if $g'(0)$ is small enough. The same problem is solved by firm j, and leads to the first order condition:

$$g'(\theta_j) = \theta_i(\pi^j_{11} - \pi^j_{10}) + (1 - \theta_i)(\pi^j_{01} - \pi^j_{00}). \tag{5.3}$$

By symmetry, $\pi^j_{11} - \pi^j_{10} = \pi^i_{11} - \pi^i_{01}$, and $\pi^i_{10} - \pi^i_{00} = \pi^j_{01} - \pi^j_{00}$. In the symmetric equilibrium, $\theta^*_i = \theta^*_j = \theta^*$ and therefore

$$\theta^* = \frac{g'(\theta^*) + \pi_{00} - \pi_{10}}{\pi_{11} + \pi_{00} - (\pi_{01} + \pi_{10})} \tag{5.4}$$

From first order conditions it can be seen that efforts spent in R&D are *strategic substitutes* if $(\pi_{10} - \pi_{00}) > (\pi_{11} - \pi_{01})$ (*i.e.* reaction curves are downward sloping), and *strategic complements* if the inequality is reversed (*i.e.* reaction curves are upward sloping). This is depicted in Figure 5.1. The reaction curves are denoted by $R_i(\theta_j)$ for firm i and $R_j(\theta_i)$ for firm j.

As a first result I state

Proposition 5.1 *The optimal effort exerted in the unconstrained equilibrium is increasing in the achieved profit levels in case of own success and decreasing in the profit levels in case of no own success. Effort levels of the competing firms exhibit strategic substitutabilities, if $(\pi_{10} - \pi_{00}) > (\pi_{11} - \pi_{01})$, and strategic complementarities, if $(\pi_{10} - \pi_{00}) < (\pi_{11} - \pi_{01})$.*

5.4 Financial Constraints, Competition and Predation

5.4.1 The Optimal Incentive Contract

The investor makes a *take it or leave it*-offer to firm i. The contract can only be based on variables that are verifiable by the investor. Since neither exerted effort nor

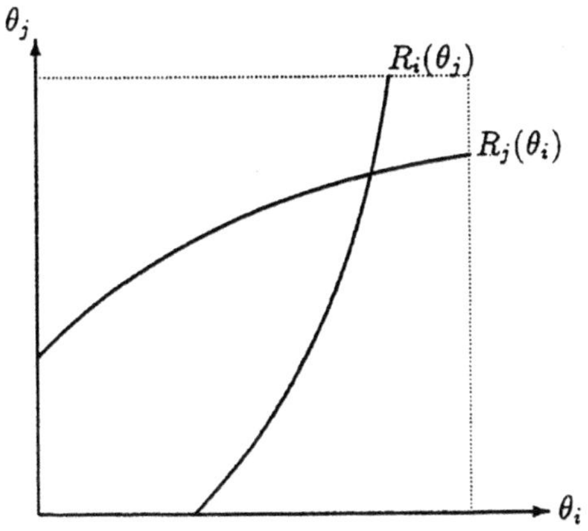

Figure 5.1 (a):
Strategic Complements

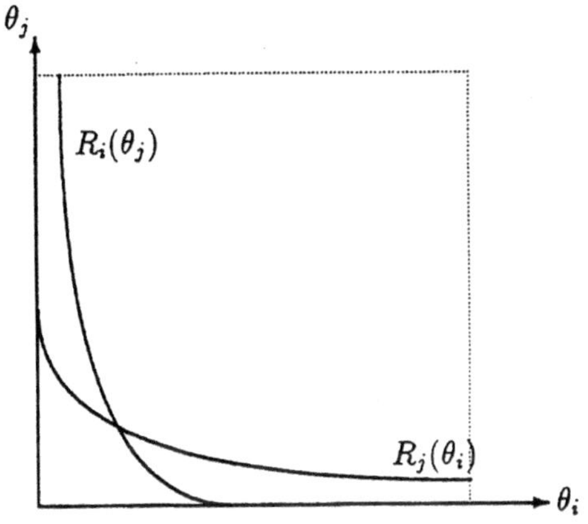

Figure 5.1 (b):
Strategic Substitutes

Figure 5.1: Equilibrium Efforts: No Financial Constraints

realized profits are verifiable, the only possible basis for the contract are reported figures. Therefore I assume that the contract is written on reported success. At the end of the period, firm i makes an announcement to the investor about the profit it realized.

Call $R_{k,l}$, $k, l = 0, 1$ the first period repayment specified by the contract if reported first period profits are $\pi_{k,l}$. The only stick the investor can choose to force firm i to exert effort in an optimal contract is the threat not to finance the firm in the second period. I therefore specify $\beta_{k,l}$ to be the probability that further finance is provided given that reported profits are π_{ij}. I assume that the firm and the investor can agree on a verifiable random device in case that the optimal probability for future finance is strictly between zero and one.

In addition, the contract specifies the repayment obligation of firm i for the second period (in the case that there is second period finance). Assume for the moment that this is π_{01} , independent of first period reports. I show later that this is indeed a part of an optimal contract.

The Role of Information Structure

Two possible information structures can be distinguished:

First, consider that the lender publicly announces a contract and firm i decides whether to sign the contract or not. After the contract is signed, both firms decide on their effort rates. If the contract is announced before the choice of effort rates the game between the lender and firm j corresponds to a *Stackelberg* game with the investor as Stackelberg leader. This case is discussed in Section 5.4.4.

Second, take the case where the contract between the lender and firm i is not publicly announced. In this case the relation between the lender and firm j can be interpreted as a *simultaneous move* game. The lender takes the effort choice of firm j as given, and offers the optimal incentive contract derived in the last section. Firm j cannot directly observe the contract, but can rationally infer that an optimal incentive contract has been signed.[8] I treat this case first.

Derivation of the Contract

The optimal incentive contract is a list $\{R_{k,l}, \beta_{k,l}, \pi_{01}\}$, $k, l = 0, 1$ that maximizes the expected profit of the investor under the conditions that the firm truthfully reveals its profits and the participation constraints of both sides are fulfilled. This means that we are in the case where the investor is actually willing to offer the contract (the expected repayments exceed the expected investment specified in the contract), and that the firm's limited liability constraint is fulfilled.[9]

I look for a subgame-perfect equilibrium. Firms choose effort levels after the contract has been signed. The investor therefore offers an contract that is optimal for him, given that equilibrium effort levels are chosen thereafter.

Given that a contract has been signed, the equilibrium effort level of firm i solves

$$\text{(E)} \quad \begin{aligned} g'(\theta_i) = {} & \theta_j(\pi_{11} - R_{11} - \pi_{01} + R_{01} + (\beta_{11} - \beta_{01})(\bar{\pi} - \pi_{01})) \\ & +(1-\theta_j)(\pi_{10} - R_{10} - \pi_{00} + R_{00} + (\beta_{10} - \beta_{00})(\bar{\pi} - \pi_{01})). \end{aligned} \tag{5.5}$$

[8]Although firm j cannot observe the contract, it may know that i has a shallow pocket and conclude that i relies on some sort of outside finance.

[9]If the firm is protected against negative profits in all states of the world, it consequently earns positive expected profits. The participation constraint is therefore non-binding in the case of limited liability.

This is the first-order condition analogous to equation (5.2) in the case of financial constraints.

Let R be the expected profit of the investor. The investor maximizes its expected revenue R through the choice of $\{R_{k,l}, \beta_{k,l}\}$, $k, l = 0, 1$ and therefore solves

$$\max_{R_{ij},\beta_{ij}} \quad \theta_i\theta_j R_{11} + \theta_i(1-\theta_j)R_{10} + (1-\theta_i)\theta_j R_{01} \qquad (5.6)$$
$$+(1-\theta_i)(1-\theta_j)R_{00} - F + (\beta_{11}\theta_i\theta_j + \beta_{10}\theta_i(1-\theta_j)$$
$$+\beta_{01}(1-\theta_i)\theta_j + \beta_{00}(1-\theta_i)(1-\theta_j))(\pi_{01} - F) := R$$

subject to

$$\text{(E)} \quad g'(\theta_i) = \theta_j(\pi_{11} - R_{11} - \pi_{01} + R_{01} + (\beta_{11} - \beta_{01})(\bar{\pi} - \pi_{01}))$$
$$+(1-\theta_j)(\pi_{10} - R_{10} - \pi_{00} + R_{00} + (\beta_{10} - \beta_{00})(\bar{\pi} - \pi_{01}))$$
$$\text{(IR(F))} \quad R_{ij} \le \pi_{ij}$$
$$\text{(IR(I))} \quad R \ge 0$$
$$\text{(IC 1)} \quad \pi_{10} - R_{10} + (\beta_{10} - \beta_{11})(\bar{\pi} - \pi_{01}) \ge \pi_{10} - R_{11}$$
$$\text{(IC 2)} \quad \pi_{10} - R_{10} + (\beta_{10} - \beta_{00})(\bar{\pi} - \pi_{01}) \ge \pi_{10} - R_{00}$$
$$\text{(IC 3)} \quad \pi_{10} - R_{10} + (\beta_{10} - \beta_{01})(\bar{\pi} - \pi_{01}) \ge \pi_{10} - R_{01}$$
$$\text{(IC 4)} \quad \pi_{11} - R_{11} + (\beta_{11} - \beta_{00})(\bar{\pi} - \pi_{01}) \ge \pi_{11} - R_{00}$$
$$\text{(IC 5)} \quad \pi_{11} - R_{11} + (\beta_{11} - \beta_{01})(\bar{\pi} - \pi_{01}) \ge \pi_{11} - R_{01}$$
$$\text{(IC 6)} \quad \pi_{00} - R_{00} + (\beta_{00} - \beta_{01})(\bar{\pi} - \pi_{01}) \ge \pi_{00} - R_{01}.$$

The objective function consists of the expected repayment net of the operating cost taken by the investor plus the second period repayment multiplied by the total probability of future finance.

The first constraint (E) reflects the implications the contract might have on the effort chosen by the firm. It is the first order condition of the leveraged Firm i obeying the conditions of the contract.

The two (IR)-constraints for the firm and the investor guarantee that the contract is actually offered and accepted and that the firms limited liability constraint is fulfilled, given that second period repayment conditional on finance in both periods is π_{01}.

The (IC)-constraints guarantee truthful report of profits. The solution to the problem of the investor is summarized in the following theorem:

Theorem 5.1 *An optimal incentive contract in the case with serially uncorrelated profits induces optimal effort for the leveraged firm. It can be characterized by a list of repayment obligations* $\bar{R} = (\bar{\pi}, \bar{\pi}, \pi_{00}, \pi_{01})$*, and a list of probabilities for second period finance* $\bar{\beta} = (1, 1, \frac{\pi_{00}-\pi_{01}}{\bar{\pi}-\pi_{01}}, 0)$*. In the second period the firm pays back* π_{01}*, irrespectively of which profit it realized. This contract is only offered if the expected revenue to the investor is positive, which is given by*

$$\begin{aligned} R &= \theta_i(\bar{\pi} + \pi_{01} - F) + \theta_j(1 - \theta_i)\pi_{01} \\ &\quad + (1 - \theta_i)(1 - \theta_j)(\pi_{00} + \frac{\pi_{00} - \pi_{01}}{\bar{\pi} - \pi_{01}}(\pi_{01} - F)) - F. \end{aligned}$$

Proof: See Appendix A.6. I only show here that second period repayment is π_{01} in the optimal contract. Note that the investor cannot extract a higher repayment in the second period in any case. Call second period repayment R_2. If reported profits are π_{00}, the repayment obligation cannot be increased. By (IC 6) binding this leads to an increase in $\beta_{00} = \frac{\pi_{00}-\pi_{01}}{\bar{\pi}-R_2}$. By (IC 1) and (IC 4) I have

$$R_{10} = R_{11} = R_{00} + (1 - \beta_{00})(\bar{\pi} - R_2) = \bar{\pi} - \pi_{01} - R_2.$$

A decrease in R_2 therefore leads to an increase in repayment of the same size if profits are high. Since this happens with probability smaller than one, there is a net loss from lowering R_2. □

Discussion: The contract derived here does not distort the incentives for the effort choice of the leveraged firm, but provides insufficient finance for the firm in the second period.[10] Although expected net profits are positive in both periods, the firm will only find finance if first period profits are not too low. Note that this outcome is inefficient, since the expected profit from further finance is always positive. Nevertheless there is no scope for renegotiation between the leveraged firm and the lender. The lender cannot hope that he will have returned his money after the second period, since there is no possible contract that enforces truthful report of second-period profits.

The result hinges on the assumption that profits are not directly observable by the lender. Suppose instead that they are observable. In this case there would always

[10]This result is dependent on the specification of the timing and the unobservable choice variables of the firm. One must not expect undistorted effort in the case where the firm can in addition choose the type (i.e. the riskiness) of the project (Bester and Hellwig [1987] provide an example of this sort).

be finance in the second period, since every innovation project is *ex ante* profitable. Without any further commitment device, any contract that specifies no finance for the second period in some cases, would be renegotiated.

In the present analysis it was assumed that the financed firm does not use first period surplus to finance second period operating costs. This assumption might look a bit artificial, but does not alter the results as long as the participation constraint of the investor is not binding.[11]

The optimal incentive contract derived in this section resembles a standard debt contract. The firm is obliged to pay back a certain amount $\bar{\pi}$ if successful. If the firm fails all the money it has earned goes to the bank (or lender) and the bank refuses to further finance the firm (or provides finance only with some probability strictly smaller than one).

In Section 5.2 I summarized the literature on optimal debt contracts. It turned out that standard debt is optimal under quite general conditions. It can therefore be conjectured that the optimality of a contract like the one derived here is preserved under a wide range of alternative specifications.

5.4.2 Competition and Equilibrium Predation

In the last section the optimal contract between an investor and a firm with a shallow pocket was derived. When both firms are able to finance their projects themselves the outcome is a symmetric equilibrium with identical effort rates in both periods.

When Firm i is dependent on outside finance, an additional effect has to be taken into account. Since the optimal contract is only providing finance in the second period if first period reports were high, a rival firm may want to reduce the probability of the occurrence of high profits.

The aggregate probability of second period finance for Firm i under the optimal

[11] The firm can only retain earnings for future innovation projects in the case of high profits. But, in the case of high profits, the outside claimholders always continues funding when the contract is signed. If the participation constraint is binding, it can be relaxed by retained equity in case of high profit when the contract has the provision of maximum equity participation (See Gale and Hellwig [1985] on the role of equity participation).

incentive contract is

$$Prob(\text{finance in } t = 2) = \theta_i + (1 - \theta_i)(1 - \theta_j)\frac{\pi_{00} - \pi_{01}}{\bar{\pi} - \pi_{01}}. \tag{5.7}$$

An increase in effort for Firm j leads to a decreased probability in second period finance of the other firm, or

$$\frac{\partial Prob(\text{finance in } t = 2)}{\partial \theta_j} = -(1 - \theta_i)\frac{\pi_{00} - \pi_{01}}{\bar{\pi} - \pi_{01}}. \tag{5.8}$$

Let $\bar{\pi}_m^j$ be the expected monopoly profit of Firm j in the second period. Firm j then chooses its first period effort according to the new first order condition

$$g'(\theta_j) = \theta_i(\pi_{11}^j - \pi_{10}^j) + (1 - \theta_i)(\pi_{01}^j - \pi_{00}^j + \frac{\pi_{00} - \pi_{01}}{\bar{\pi} - \pi_{01}}(\bar{\pi}_m^j - \bar{\pi}^j)). \tag{5.9}$$

As a result, the deep pocket firm j chooses a higher first-period effort in equilibrium when there is a leveraged rival. The leveraged firm reacts to this increase in rivals effort depending on whether effort rates exhibit strategic substitutabilities or complementarities. Therefore the following theorem can be stated:

Theorem 5.2 *If one firm is dependent on outside finance, rivals choose to exert higher effort in the first period compared to the symmetric case without outside finance. The leveraged firm itself chooses higher first period effort rates if innovational efforts of the firms are strategic complements* $((\pi_{10} - \pi_{00}) < (\pi_{11} - \pi_{01}))$, *and decreases effort in the case of strategic substitutes* $((\pi_{10} - \pi_{00}) > (\pi_{11} - \pi_{01}))$.

The result is illustrated in figure 5.2.

Discussion: The predation result just derived contrasts the findings of Brander and Lewis. A firm that chooses to finance its innovation projects by outside debt might find itself in a weaker situation in the product market if rivals take advantage of the opportunity to prey. There might also be an increase in the indebted firm's effort, but only in specific strategic environments.

Note that the result does not depend on the contract being an optimal incentive contract. Every standard long-term debt contract that specifies continuation of finance under conditions that can be influenced by rivals creates incentives to prey.

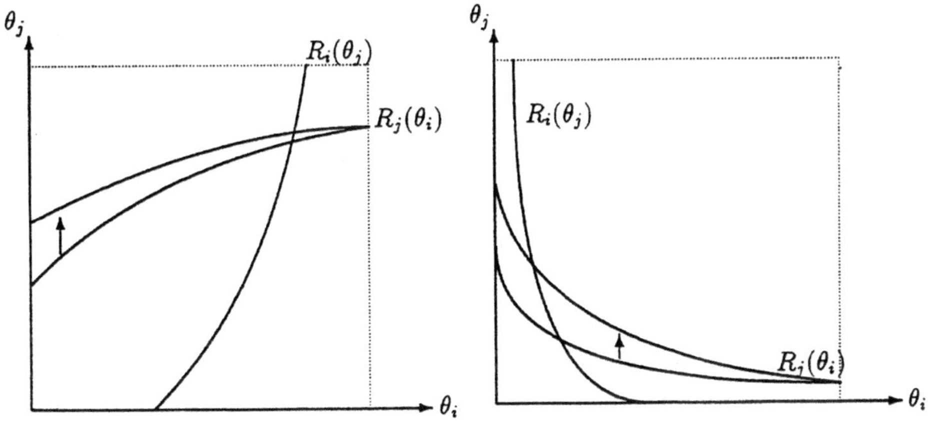

Figure 5.2 (a):
Strategic Complements

Figure 5.2 (b):
Strategic Substitutes

Figure 5.2: Equilibrium Efforts: The Predation Effect

I already stressed the point that, although the contract is derived as an optimal contract in a restricted setting, the optimality extents to a variety of alternative situations provided that there is ex post informational asymmetry between the lender and the firm.

Entry Deterrence by an Incumbent with a Reputation to be "Tough"

There is yet another possibility I have neglected so far. Suppose the strategic situation is changed a bit. Think of the deep pocket firm j to be an incumbent, threatened by the potential (outside financed) entry of a rival. The incumbent has built a reputation for being "tough" or is able to commit himself (through some strategic device or simply as a repeated-game effect) to a higher effort. According to equation (6) the financier of the entrant i receives an expected payoff of

$$\begin{aligned} R &= \theta_i(\bar{\pi} + \pi_{01} - F) + \theta_j(1-\theta_i)\pi_{01} \\ &\quad +(1-\theta_i)(1-\theta_j)(\pi_{00} + \frac{\pi_{00}-\pi_{01}}{\bar{\pi}-\pi_{01}}(\pi_{01}-F)) - F. \end{aligned}$$

Ignoring the second order effect on the leveraged firms' choice of effort[12], the impact

[12]This effect is of ambiguous sign, depending on the slopes of the reaction curves, i.e., whether

of an increase in the incumbent's effort is

$$\frac{\partial R}{\partial \theta_j} = -(1-\theta_i)(\pi_{00} + \frac{\pi_{00}-\pi_{01}}{\bar{\pi}-\pi_{01}}(\pi_{01}-F) - \pi_{01}) \tag{5.10}$$

This can be re-written as

$$-(1-\theta_i)\left(\frac{(\pi_{00}-\pi_{01})(\bar{\pi}-F)}{\bar{\pi}-\pi_{01}}\right) < 0 \tag{5.11}$$

If the investor's profit is not too high, the incumbent find it profitable to deter entry by threatening the potential entrant with an first period effort θ_j' high enough to make the investor's expected profit negative. In this case, an entrant with a shallow pocket will not find finance for his project and the incumbent's monopoly position remains unchallenged.

5.4.3 An Example

The firms are competing in the product market, facing a linear inverse demand function $p = 6 - x_1 - x_2$. If they are successful with an innovation in one period, they have constant marginal costs of production $\underline{c} = 1$; if not, $\bar{c} = 2$. The firms play the Cournot-Nash equilibrium in the product market. The profit matrix for firm i is thus

$i \backslash j$	success	no success
success	$\frac{25}{9}$	$\frac{36}{9}$
no success	1	$\frac{16}{9}$

The effort function g is assumed to be assumed quadratic, i.e. $g = 2\theta^2$. The optimal success probability in the symmetric unconstrained problem can be calculated following equation (5.2). The result is $\theta_i^* = \theta_j^* = \theta^* = \frac{1}{2}$, $g(\theta^*) = \frac{1}{2}$. Under the optimal incentive contract, $\beta_{00} = \frac{14}{25}$.

efforts are strategic substitutes or complements. On the other hand, the effect is relatively "small". I give an example in the next section.

Since optimal effort is induced, the predation equilibrium (θ_i', θ_j') solves the equations (5.2) and (5.9).[13] The expected monopoly profit of Firm j net of effort is approximately 4.63.Thus, I have

$$4\theta_i = \frac{20}{9} - \frac{4}{9}\theta_j$$

and

$$4\theta_j \approx 3.76 - 1.98\theta_i$$

which is solved by

$$\theta_i' \approx 0.477, \qquad \theta_j' \approx 0.688.$$

This is the predation effect predicted by the model: The leveraged firm reduces its effort, while the rival increases its effort. The example with Cournot duopoly is the standard case with strategic substitutes, where the two firms react in different directions to the changed strategic situation. The expected profit of the investor can be calculated from equation (5.7). This yields

$$R(\theta_i', \theta_j') = 1.02 - 1.568F.$$

Thus, as long as $F < 0.65$ the contract will actually be offered by the investor. Suppose $F = 0.6$ and Firm j is able to commit to $\hat{\theta}_j = 1$ in the first period. The optimal choice of Firm i in reaction to the commitment of Firm j would be $\hat{\theta}_i = \frac{4}{9}$. The expected profit of the investor is then

$$R(\hat{\theta}_i, \hat{\theta}_j) = 0.86 - \frac{13}{9}F = -0.0066$$

and Firm i will not find any finance as long as it cannot in turn commit to an higher effort itself.

[13]If optimal effort is induced, the equations (5.2) and (5.5) coincide. Equation (5.9) is the optimal effort of firm j given that it faces a leveraged rival.

5.4.4 Precommitment through Observable Contracts

In the beginning of the equilibrium analysis I described two possible specifications of the information structure of the game: The simultaneous move game discussed above and the Stackelberg game, where the bank publicly announces the type of contract she is going to offer. In this case the bank can precommit to a contract that is less sensitive to predatory actions taken by rivals.[14]

The easiest way to immunize the contract against predatory action taken by rivals is to condition the provision of future funding on the success of the financed firm. This can be achieved by reducing the probability of further finance in the case of no success of either firm (β_{00}) to zero.[15] The direct cost of reducing this probability stems from the fact that the new contract lacks full incentive compatibility. The firm always reports the lowest profit if either π_{01} or π_{00} is realized.

Let me assume that the new contract specifies that in case of success of firm i the bank provides finance in the second period, while no finance is provided if i fails. The repayment in the first period will then be $\bar{\pi}$ if reported profits are high, while π_{01} otherwise. Remember Condition (E) from equation (5.5):

$$\begin{aligned} g'(\theta_i) &= \theta_j(\pi_{11} - R_{11} - \pi_{01} + R_{01} + (\beta_{11} - \beta_{01})(\bar{\pi} - \pi_{01})) \\ &\quad +(1-\theta_j)(\pi_{10} - R_{10} - \pi_{00} + R_{00} + (\beta_{10} - \beta_{00})(\bar{\pi} - \pi_{01})) \end{aligned}$$

Plugging in the conditions of the new contract leads to

$$g'(\theta_i) = \theta_j(\pi_{11} - \pi_{01}) + (1-\theta_j)(\pi_{10} - \pi_{00}). \tag{5.12}$$

This is the same as equation (5.2), and therefore the effort choice of the financed firm i is not distorted from the change in the contract. The direct cost of the change in the contract is given by

$$(\pi_{00} - \pi_{01})(1 - \frac{F - \pi_{01}}{\bar{\pi} - \pi_{01}}) > 0. \tag{5.13}$$

[14] I implicitly assume here that there is only precommitment in the case of observable contracts. This is true under the given structure of the game. For alternative settings, where unobservable contracts may also serve as a commitment device, see Katz [1991].

[15] Alternatively, the contract may specify any value for β_{00} and β_{01} as long as they are identical, but in this case it is optimal to choose the lowest possible values.

This direct loss has to be weighted against the gain from inhibiting predation by rivals. Thus, as in the model of Bolton and Scharfstein [1990] it may be profitable for the lender to announce a contract that renders predation impossible. The result is summarized in the next proposition:

Proposition 5.2 *There is an optimal no-predation contract. This contract provides finance only in the case of success of the leveraged firm i. As before, the effort choice of firm i is not distorted.*

Proof: It remains to be shown that the no-predation contract derived above is indeed optimal. The proof is similar to the proof of the optimal contract above. Therefore I give only a sketch of the proof: The bank cannot extract more than $\bar{\pi}$ when profits are high without violating (IC 2) and (IC 4). At the same time it is not profitable to choose a probability of second period finance strictly lower than one in case of high profits as long as $\bar{\pi} > F$ i.e., as long as the investment is profitable in expected terms. Therefore the terms of the optimal contract are unchanged for the case in which firm i succeeds in the first period. In the case of failure in the first period, a marginal increase in the probability of finance if reported profits are π_{01} induces a marginal loss of $F - \pi_{01}$, which is bigger than the expected marginal loss from reducing β_{00} (and therefore reducing R_{00}) from Lemma A.3 in the proof of Theorem 5.1. Therefore the proposed no-predation contract is indeed optimal. □

Remark: Though the new contract renders predation impossible, and thus works as a commitment device, there is even greater inefficiency than under a contract without commitment, since now the financed firm i never gets finance for the second period unless it succeeds in the first period.

Still another point is the possibility of renegotiation of the contract. Suppose that firm i realized π_{00}. The bank cannot force the firm to report this profit, but it is in the interest of the firm to offer renegotiation of the contract. Since the bank gains from renegotiation (see equation 5.13), and so does firm i, renegotiation is likely to occur. I therefore state the as a final result:

Theorem 5.3 *The optimal no-predation contract described in Proposition 5.2 is not renegotiation-proof.*

Discussion: Is renegotiation likely to occur? – Note that I have assumed that renegotiation is costless. In a world where this assumption is not valid, renegotiation would only occur if the possible gains of the investor are greater than the transaction costs generated by renegotiation.

A further objection is that at least banks could implement commitment devices to prevent them from renegotiation. One such commitment device is the development of simple rules how credits have to be handled by the employees in standard cases and the invention of complicated procedures for exceptions from that rule. It is therefore an empirical question whether such arrangements are actually used.

A last objection is that the theorem above is derived from a highly specific model. Bolton and Scharfstein [1990] show that in their framework renegotiation does not occur. It is therefore up to future investigation whether their model or the specification used here is more robust concerning the renegotiation proofness of a no-predation contract.

5.5 Summary and Conclusion

I analyzed a simple multi-period innovation model with market competition between two firms. Firms choose efforts to complete research projects. To start a project, a fixed cost has to be incurred. One firm is financially constrained (i.e. it has insufficient equity and owns no assets that are collaterizable) while the other firm is not. The financially constrained firm writes an optimal incentive contract with an outside investor.

I analyzed two alternative informational structures:

First, the finance contract between the outside investor and the firm that needs outside funding is unobservable, but can rationally be inferred by the competitor. In the *second* variant, I allowed the bank to precommit to a contract that renders impossible predation by the rival.

I summarize the main results:

- The optimal contract between the investor and a firm resembles a *standard debt contract.* It specifies a fixed repayment obligation. If the firm cannot pay the agreed repayment the whole obtained profit goes to the bank. Future finance is then provided with probability lower than one. Thus, there is insufficient finance despite the fact that the second period project is profitable in expected terms.

- The optimal contract enforces truthful revelation of profits without distorting the effort choice of the firm.

- Under the first information structure (no announcement of the contract), the leveraged firm is vulnerable by rival attacks. The unleveraged rivals increases its first period efforts, and thereby decreases the probability of future finance for the leveraged firm. Thus, the unleveraged rival takes *predatory* action to drive the leveraged firm out of the market. This behavior is in accordance with empirical findings by Kovenock and Phillips [1993, 1995].

- There is an optimal no-predation contract under the second information structure. This contract provides finance at an even lower probability in the second period than is the case in the optimal contract that allows for predation. There is thus an additional loss in efficiency. This is also the reason why this contract is not renegotiation-proof. This result is opposed to the results of Bolton and Scharfstein.

The predation result derived here is especially relevant for young and small innovative firms that do not possess collateral or retained earnings to allow for internal funding or eased access to capital markets.

In the light of the findings of Levin *e.a.* [1987] that in most industries institutional means of rent protection (i.e. patents) is viewed as being ineffective, one should expect established firms using the opportunities to prey innovative but undercapitalized firms.

The behavioral implications derived in this chapter are that the rivals of leveraged firms tend to increase the efforts to innovate compared to the case without financial constraints. The behavior of leveraged firms is less clear-cut. They tend to reinforce their efforts when the potential losses from failing are increasing when rivals succeed (which is the case of strategic complementarities in innovative effort levels). When the potential losses decrease, the optimal reaction to the higher efforts of rivals is to decrease own innovative efforts.

I conclude with some critical comments on the framework chosen above:

The notion of *innovative efforts* chosen in the present analysis is quite vague and may be given a number of different interpretations. A possible interpretation is the use of non-collaterizable assets such as the *human capital* incorporated in the firm. In the present model innovative effort was introduced as a private cost (disutility). This is not a necessary assumption. Think about physical investment into assets to conduct R&D as being the fixed cost of innovation and additionally variable costs varying with the intensity with which the assets are used. The assets cannot serve as collateral for outside financiers since they have only very specific usages, and are therefore much less valued by outsiders compared to the value the firm attaches to those assets.

However, I conjecture that the qualitative results of the analysis do not change if variable costs are included into the limited liability constraints of the firm.

Another point is the rigid structure of the present approach. In the light of the discussion in the previous chapters, a dynamic model with sequential entry and the choice of innovation quality under financial constraints would be rewarding extension to the present analysis.

The last point is the robustness of the result that the no-predation contract is not renegotiation proof. This point remains an open topic for future research.

Chapter 6

Summary and Conclusions

The broad theme of this dissertation is *technological competition*. As defined in the Introduction this term denotes the interaction between industrial structure and innovative activity by firms. From the review of the literature in Chapter 2 it is clear that there is a diversity of approaches.

Facing diversity, the best thing to do is to search for a rule to tell the differences. What determines the comparative statics of game-theoretical models used to analyze industrial structures? - Reading the articles by Milgrom and Shannon [1994], Bulow, Geanakoplos and Klemperer [1985], and Fudenberg and Tirole [1984] served me to give at least a partial answer to this question. It is what I called the *strategic structure* of the game, namely whether the strategies chosen by firms are substitutes or complements for each other.

In Chapter 3 I have shown that there is a close connection between the division of profits between firms in the post-innovation market and the incentives to innovate. The strategic structure of the first innovation stage game generates the link: In a game of quality choice (of the innovation) the firms' equilibrium strategies are complements in the case of first-mover advantages, while they are substitutes in the case of second-mover advantages.

The situation is sometimes complicated by the fact that the strategic structure is not exogenous, but endogenously determined through the coordination of firms on a specific equilibrium in the presence of multiple equilibria.

In Chapter 4 I analyzed the timing of the commercial use of innovation. Is there a case to be made in which firms decide to postpone the introduction of an innovation into a market after completion of the research stage? - I showed that this is indeed the case if an additional source of uncertainty about the "productiveness" of the research approach is introduced (*hazard rate uncertainty*), and if the research approaches of competing firms are identical (or at least positively correlated).

Whether this assumption is too strong cannot be answered by theory, but asks for empirical investigation. The more the research is fundamental or basic research, and the more the field in which competing firms conduct research is a narrow one, the more firms will make use of secrecy and delay.

The strategy to keep certain findings secret may serve as a powerful device to prey a rival that is financially constrained out of competition. Assume that the investors of the constrained firm do not know the details of the research program, but think that successes in the field are positively correlated. Their subjective assessment of the success probability of their debtor may increase due to successes announced in the field, and so will their willingness to continue finance. Holding back information about success may therefore hurt the financially constrained firm. The formal analysis of the interaction of timing decisions and financial constraints is not done here, and remains an interesting topic open for future research.

In Chapter 5 I analyzed the implications of financial constraints on the incentives of firms to innovate. I confirmed (and strengthened) the result of Bolton and Scharfstein [1990] that a firm that is financially constrained through an (optimal) debt contract with renewal conditioned on repayment is vulnerable to predatory actions taken by rivals.

In addition I have shown that the incentives to innovate are shaped by the strategic structure of the underlying market game. The strategic structure also determines the comparative statics of the model (with two firms): If the efforts spent on innovative activity are strategic complements, then both firms increase their efforts when one firm is dependent on outside finance compared to the case with no outside finance. If the efforts spent are strategic substitutes, then the unleveraged rival increases its effort, while the effort of the leveraged firm decreases in equilibrium.

The dissertation contains no empirical investigation, but I tried to motivate and feed the analysis with casual evidence and examples from industry studies. And it is questionable whether strategic models from industrial organization theory work well as an input for empirical studies. I am convinced that they work well as a clarifying device for the underlying structural facts. But I doubt the value of a great part of the literature to serve for implementation in structural testing of hypotheses about industrial structure and its dynamics.

Building IO-models adds to the indeterminacy of the policy instructions that can be generated from theoretical economic reasoning. On one side I contributed partly to this indeterminacy by adding three more models to the literature. On the other side I tried to emphasize the structural components of game-theoretical models that are determinants of the comparative static behavior. This may serve as a clarifying device to organize the different models and results, and therefore potentially reduces the indeterminacy of the results.

But there are still some general lessons that can be learned from the analysis.

First, full appropriability of the rents from innovation is not always beneficial. The evolution of an industry through the entry of firms with different innovations is shaped by the relative size of profits that firms expect from innovation. Even in the case where later entrants into a market earn more than the "pioneer" the incentives to innovate are not necessarily spoiled. What is true is that the evolution of the industry is strongly influenced by the the legal institutions for rent protection from innovation. And stronger appropriability of returns (by increasing first mover profits) may lead to less overall innovation and to a slowdown in the evolution of innovation quality.

Second, there may be incentives for firms to hold back information generated during the process of innovation, and these incentives are likely to be detrimental to welfare. There should therefore be institutions that create incentives to make research results public.

A *final* point is the strategic use of market power (*e.g.* a long purse) against small and innovative firms. Established firms have an incentive to take predatory actions to weaken the position of rivals which *e.g.* depend on outside funding. The reason is the asymmetry of information between firms conducting risky research projects and their financiers. The theory predicts that market institutions will generally be unable to solve these problems. This does not necessarily mean that government should directly subsidize small and research intensive firms, but calls for securitization, certification and credit insurance from public authorities as possible means to alleviate moral hazard and adverse selection problems for private investors. Another possibility is the sponsoring or creation of institutions that serve to screen and monitor research conducting firms, and thereby generate information that reduce agency problems.

I covered only a few and dispersed topics in the field of technological competition. But the field has grown so complex that I would never claim that I am able to construct a model that deserves the title "*the* theory of technological competition". That is why I restrained myself to the choice of a few interesting topics, and tried to tell a story that serves the purpose: A (weakly) negative balance between reducing and adding indeterminacy to the theory of technological competition.

Appendix

A.1 Proof of Theorem 3.1

Theorem 3.1 *Let Assumptions 3.1–3.4 be fulfilled. Then there exists a subgame perfect equilibrium for the sequential entry game with two firms.*

Proof of Theorem 3.1:

Given the quality levels realized by both firms after entry there exists an equilibrium in the market competition stage. Profits are bounded, and therefore the entry problem of the second mover has a solution. To show that there exist a subgame perfect equilibrium in the sequential entry game it suffices to show that there exists an equilibrium for the first entry stage.

Firms choose thresholds $\bar{\gamma}^i \in [0, \infty)$ as first entry strategies. The strategy space of the game is thus the set $\{(\bar{\gamma}^i, \bar{\gamma}^j) | \bar{\gamma}^i \in [0, \infty) \text{ and } \bar{\gamma}^j \in [0, \infty)\}$. An equilibrium in this game is defined as a fixed point of some mapping

$$BR : [0, \infty)^2 \longrightarrow [0, \infty)^2 \tag{A.1}$$

Consider first the case of *strategic complements.* From Proposition 3.5 BR is an increasing function on the strategy space. Take some compact subset $[a, b]$ of the strategy space such that $a \leq BR(a) \leq BR(b) \leq b$. This set forms a *complete lattice* if endowed with the product order $\geq$.[1] From Tarski's Theorem (Tarski [1955], Theorem 1) we know that every increasing function on a complete lattice has a fixed point.

It remains to show that a set $[a, b]$ with $a \leq BR(a) \leq BR(b) \leq b$ does exist. It is profitable for the firms to start research from Assumption 3.1. Thus, we have

[1]A set that is compact in any topology at least as fine as the interval topology endowed with any reflexive, antisymmetric and transitive order is a complete lattice. See e.g. Vives [1990].

$(0,0) \leq BR(0,0)$. Since I assumed that profits are bounded, there exists a quality level γ^{i*} such that for any choice of firm j, firm i prefers to enter the market after the realization of a quality level $\gamma^i \geq \gamma^{i*}$ and thus, $(\gamma^{i*}, \gamma^{j*}) \leq BR(\gamma^{i*}, \gamma^{j*})$. Therefore an equilibrium exists.

The case with ***strategic substitutes*** is completely analogous, whith the only execption that the order on the strategy set of one player has to be reverted.

It remains to proof the existence of equilibrium in the case of mixed games as asserted in Theorem 3.3. A mixed game is a game with strategic substitutabilities for low threshold levels and with strategic complementarities thereafter. Suppose there is no equilibrium in the subset of the strategy space characterized by second-mover advantages for at least one firm. In the complement of this subset the game is a game with strategic complementarities.

Let the lowest threshold levels in this region be denoted $(\hat{\gamma}^i, \hat{\gamma}^j)$. Since there is no equilibrium in a lower region of the strategy set it must be that $(\hat{\gamma}^i, \hat{\gamma}^j) \leq BR(\hat{\gamma}^i, \hat{\gamma}^j)$. To understand this assertion, suppose that the reverse holds. Again, from Assumption 3.1 we have $(0,0) \leq BR(0,0)$, and $(\hat{\gamma}^i, \hat{\gamma}^j) > BR(\hat{\gamma}^i, \hat{\gamma}^j)$. Since the set $[0, \hat{\gamma}^i] \times [0, \hat{\gamma}^j]$ is compact it follows that BR has a fixed point on this set. This leads to a contradiction since, by assumption, no equilibrium exists in the lower region.

From the boundedness assumption on expected profits we may again conclude that there exists a pair of thresholds $(\gamma^{i*}, \gamma^{j*}) \geq BR(\gamma^{i*}, \gamma^{j*})$. Thus, an equilibrium exists.□

A.2 Proof of Theorem 3.2

Theorem 3.2 *(i) Suppose that $W > L$, and the two firms are identical. Only symmetric equilibria exist. (ii) Suppose that there exists a symmetric equilibrium $\bar{\gamma}$ with $W(\bar{\gamma}) < L(\bar{\gamma})$. Then there is no other symmetric equilibrium $\hat{\gamma}$ with this property.*

Proof of Theorem 3.2: (i) The proof will be by contradiction. Assume that there exist a pair (ζ^i, ζ^j) that forms an equilibrium in the first entry game with $\zeta^i > \zeta^j$.

From (3.19) we have

$$rW(\zeta^i) - \lambda\left(\int_{\zeta^i}^{\infty}[W(\gamma^i) - W(\zeta^i)]f(\gamma^i)d\gamma^i + \int_{\zeta^j}^{\infty}[L(\gamma^j) - W(\zeta^i)]f(\gamma^j)d\gamma^j\right) =$$
$$rW(\zeta^j) - \lambda\left(\int_{\zeta^j}^{\infty}[W(\gamma^j) - W(\zeta^j)]f(\gamma^j)d\gamma^j + \int_{\zeta^i}^{\infty}[L(\gamma^i) - W(\zeta^j)]f(\gamma^i)d\gamma^i\right).$$

Re-arranging terms leads to

$$\begin{aligned}
0 = \quad & (r + \lambda\sum_{i=1}^{2}(1 - f(\zeta^i)))[W(\zeta^i) - W(\zeta^j)] \\
- \quad & \lambda\left(\int_{\zeta^i}^{\infty} W(\gamma^i)f(\gamma^i)d\gamma^i - \int_{\zeta^j}^{\infty} W(\gamma^j)f(\gamma^j)d\gamma^j\right) \\
- \quad & \lambda\left(\int_{\zeta^j}^{\infty} L(\gamma^j)f(\gamma^j)d\gamma^j - \int_{\zeta^i}^{\infty} L(\gamma^i)f(\gamma^i)d\gamma^i\right) \\
> \quad & \lambda\left(\int_{\zeta^j}^{\infty} W(\gamma^j)f(\gamma^j)d\gamma^j - \int_{\zeta^i}^{\infty} W(\gamma^i)f(\gamma^i)d\gamma^i\right) \\
+ \quad & \lambda\left(\int \zeta^{i\infty} L(\gamma^i)f(\gamma^i)d\gamma^i - \int_{\zeta^j}^{\infty} L(\gamma^j)f(\gamma^j)d\gamma^j\right) \\
= \quad & \lambda\left(\int_{\zeta^j}^{\zeta^i} W(\gamma)f(\gamma)d\gamma - \int_{\zeta^j}^{\zeta^i} L(\gamma)f(\gamma)d\gamma\right) \\
= \quad & \lambda\int_{\zeta^j}^{\zeta^i}[W(\gamma) - L(\gamma)]f(\gamma)d\gamma,
\end{aligned}$$

where the inequality comes from Assumption 3.3. Thus, we have

$$0 > \lambda\int_{\zeta^j}^{\zeta^i}[W(\gamma) - L(\gamma)]f(\gamma)d\gamma,$$

which leads to a contradiction because $W > L$ by assumption.

(ii) Suppose that there are two symmetric equilibria $\hat{\gamma}, \bar{\gamma}$ such that $\hat{\gamma} > \bar{\gamma}$. It follows that

$$W(\hat{\gamma}) \geq W(\bar{\gamma})$$

from Assumption 3.4. Furthermore,

$$rW(\hat{\gamma}) = -c + \lambda\int_{\hat{\gamma}}(W(\gamma) + L(\gamma) - 2W(\hat{\gamma}))dF(\gamma)$$

and

$$rW(\bar{\gamma}) = -c + \lambda \int_{\bar{\gamma}} (W(\gamma) + L(\gamma) - 2W(\bar{\gamma}))dF(\gamma)$$

from equation (3.19). From this one gets

$$\int_{\hat{\gamma}} (W(\gamma) + L(\gamma) - 2W(\hat{\gamma}))dF(\gamma) \geq \int_{\bar{\gamma}} (W(\gamma) + L(\gamma) - 2W(\bar{\gamma}))dF(\gamma).$$

Re-arranging terms leads to

$$0 \leq 2(1 - F(\bar{\gamma}))[W(\hat{\gamma}) - W(\bar{\gamma})] \leq \int_{\bar{\gamma}}^{\hat{\gamma}} (2W(\hat{\gamma}) - W(\gamma) - L(\gamma))dF(\gamma)$$

Furthermore,

$$(1 - F(\bar{\gamma}))[W(\hat{\gamma}) - W(\bar{\gamma})] - \int_{\bar{\gamma}}^{\hat{\gamma}} (W(\hat{\gamma}) - W(\bar{\gamma}))dF(\gamma) =$$

$$(1 - F(\hat{\gamma}))[W(\hat{\gamma}) - W(\bar{\gamma})] + \int_{\bar{\gamma}}^{\hat{\gamma}} (W(\gamma) - W(\bar{\gamma}))dF(\gamma) > 0.$$

Thus,

$$0 \leq (1 - F(\bar{\gamma}))[W(\hat{\gamma}) - W(\bar{\gamma})] \leq \int_{\bar{\gamma}}^{\hat{\gamma}} (W(\hat{\gamma}) - L(\gamma))dF(\gamma)$$

or,

$$\int_{\bar{\gamma}}^{\hat{\gamma}} (W(\hat{\gamma}) - L(\gamma))dF(\gamma) \geq 0.$$

If this is the case, then $W(\hat{\gamma}) \geq L(\hat{\gamma})$ since L is decreasing in the threshold of the first entrant. Thus, at most one symmetric equilibrium with second-mover advantages exists. □

A.3 Proof of Proposition 3.10

Proposition 3.10 *Suppose that the first-entry game has multiple equilibria. The equilibria can be ranked completely with respect to the product order over the joint strategy space introduced before. There exists a highest and a lowest equilibrium, and the comparative statics results derived for the case with an unique equilibrium also hold for the highest and the lowest equilibrium in a game with multiple equilibria.*

Proof of Proposition 3.10: Restrict the strategy space to the compact subset $[0, \gamma^{i*}] \times [0, \gamma^{j*}]$ where $(\gamma^{i*}, \gamma^{j*})$ are the implicit bounds derived in the previous proof. Consider the case with *strategic complementarities.* The case with *strategic substitutabilities* is again analogous (see Appendix A.1). The best response map is denoted BR. It is increasing from Proposition 3.5.

It follows directly from Tarski's Theorem that the set of equilibria on this restricted strategy set forms a complete lattice, i.e. it contains a smallest and a largest element with respect to the product order $\geq$ on this set. Thus, a lowest and a highest equilibrium exists.

It remains to show that the monotone comparative static results hold also in the case of multiple equilibria. Note that the game with strategic complementarities analyzed here is a *supermodular game*, i.e. the expected profit of the firms have positive cross-derivatives in the threshold levels.

Assume that, from the first order condition of firm i, an increase in the value of some parameter t leads to a *ceteris paribus* increase in the best response $\bar{\gamma}^i$ (for a fixed $\bar{\gamma}^j$). This means that the expected profits of firm i exhibit *increasing differences* in $t, \bar{\gamma}^i$.

In a supermodular game, the lowest and the highest equilibrium move monotonically with a parameter t of the game if the profit functions of the players $i = 1, 2$ exhibit increasing differences in (t, γ^i). See e.g. Vives [1990], Sections 3 and 4, for a proof of this assertion. A generalization to the general class of games with strategic complementarities is found in Milgrom and Shannon [1994].□

A.4 PROOF OF THEOREM 4.2

Theorem 4.2: *The delay period $(\bar{t} - t^*)$ is shortened by an increase in the discount rate r as well as in the monopoly profits M. An increase in duopoly profits D as well as in the efficiency parameter λ^h leads to a longer delay period.*

Proof:

(i) Total differentiation of (4.19) with respect to $(\bar{t} - t^*)$ and r leads to

$$\frac{d(\bar{t} - t^*)}{dr} =$$

$$-\frac{(\bar{t}-t^*)e^{-r(\bar{t}-t^*)}[D+(M-D)e^{-\lambda^h(\bar{t}-t^*)}]+\lambda^h(M-D)}{rDe^{-r(\bar{t}-t^*)}+(r+\lambda^h)(M-D)e^{-\lambda^h(\bar{t}-t^*)}}<0.$$

(ii) Total differentiation with respect to $(\bar{t}-t^*)$ and D leads to

$$\frac{d(\bar{t}-t^*)}{dD}=\frac{e^{-r(\bar{t}-t^*)}(1-e^{-\lambda^h(\bar{t}-t^*)})-\frac{\lambda^h}{r+\lambda^h}}{rDe^{-r(\bar{t}-t^*)}+(r+\lambda^h)(M-D)e^{-\lambda^h(\bar{t}-t^*)}},$$

while the same expression with respect to M is given by

$$\frac{d(\bar{t}-t^*)}{dM}=\frac{e^{-(r+\lambda^h)(\bar{t}-t^*)}-\frac{r}{r+\lambda^h}}{rDe^{-r(\bar{t}-t^*)}+(r+\lambda^h)(M-D)e^{-\lambda^h(\bar{t}-t^*)}}.$$

From equation (4.19) it holds that

$$M(e^{-(r+\lambda^h)(\bar{t}-t^*)}-\frac{r}{r+\lambda^h})=D(\frac{\lambda^h}{r+\lambda^h}-e^{-r(\bar{t}-t^*)}(1-e^{-\lambda^h(\bar{t}-t^*)})).$$

Therefore it is clear that

$$sign\ \{e^{-(r+\lambda^h)(\bar{t}-t^*)}-\frac{r}{r+\lambda^h}\}=$$

$$-\ sign\ \{e^{-r(\bar{t}-t^*)}(1-e^{-\lambda^h(\bar{t}-t^*)})-\frac{\lambda^h}{r+\lambda^h}\},$$

and thus

$$sign\ \frac{d(\bar{t}-t^*)}{dM}=-\ sign\ \frac{d(\bar{t}-t^*)}{dD}.$$

Now suppose that

$$e^{-(r+\lambda^h)(\bar{t}-t^*)}=\frac{r}{r+\lambda^h}.$$

Substituting into equation (4.19) yields

$$\frac{rM+\lambda^h D}{r+\lambda^h}=De^{-r(\bar{t}-t^*)}+\frac{r(M-D)}{r+\lambda^h}.$$

Re-arranging terms leads to

$$D=De^{-r(\bar{t}-t^*)}$$

and thus to a contradiction as long as $\bar{t} \neq t^*$ (which is always the case). To satisfy equation (4.19) it must therefore be that

$$e^{-(r+\lambda^h)(\bar{t}-t^*)} < \frac{r}{r+\lambda^h}.$$

Consequently the delay period shrinks with an increase in monopoly profits, and therefore grows with an increase in duopoly profits.

(iii) Total differentiation with respect to $(\bar{t} - t^*)$ and λ^h gives

$$\frac{d(\bar{t}-t^*)}{d\lambda^h} = \frac{(M-D)(r - \lambda^h e^{-(r+\lambda^h)(\bar{t}-t^*)})}{rDe^{-r(\bar{t}-t^*)} + (r+\lambda^h)(M-D)e^{-\lambda^h(\bar{t}-t^*)}}.$$

The sign of this expression depends on the sign of

$$r - \lambda^h e^{-(r+\lambda^h)(\bar{t}-t^*)}.$$

From the argument above one has

$$r > (r+\lambda^h)e^{-(r+\lambda^h)(\bar{t}-t^*)} > \lambda^h e^{-(r+\lambda^h)(\bar{t}-t^*)}.$$

Therefore a higher efficiency leads to a lenghtened delay period. □

A.5 Proof of Theorem 4.3

Theorem 4.3: *The exit date $\bar{t}$ is increasing in monopoly profits M as well as in the discount rate r. It declines in the research expenditure rate c. The effect of a change in the efficiency parameter λ^h or in duopoly profits D is indeterminate. An increase in the delay period due to parameter changes (see previous theorem) ceteris paribus leads to earlier exit.*

Proof: Let me first consider the direct effects for a fixed delay period. Total differentiation with respect to the exit date $\bar{t}$ and monopoly profits M gives

$$\left.\frac{d\bar{t}}{dM}\right|_{\bar{t}-t^*=const.}$$

$$= -\frac{\lambda(\bar{t})e^{-\lambda^h(\bar{t}-t^*)}}{\lambda'(\bar{t})[D+(M-D)e^{-\lambda^h(\bar{t}-t^*)}] - \lambda^h\lambda(\bar{t})(M-D)e^{-\lambda^h(\bar{t}-t^*)}} > 0.$$

A change in duopoly profits D leads to

$$\frac{d\bar{t}}{dD}\Big|_{\bar{t}-t^*=const.}$$

$$= -\frac{\lambda(\bar{t})(1-e^{-\lambda^h(\bar{t}-t^*)})}{\lambda'(\bar{t})[D+(M-D)e^{-\lambda^h(\bar{t}-t^*)}]-\lambda^h\lambda(\bar{t})(M-D)e^{-\lambda^h(\bar{t}-t^*)}} > 0.$$

The effect of a change in research costs c is determined by

$$\frac{d\bar{t}}{dc}\Big|_{\bar{t}-t^*=const.}$$

$$= \frac{1}{\lambda'(\bar{t})[D+(M-D)e^{-\lambda^h(\bar{t}-t^*)}]-\lambda^h\lambda(\bar{t})(M-D)e^{-\lambda^h(\bar{t}-t^*)}} < 0,$$

while a change in the efficiency parameter λ^h leads to

$$\frac{d\bar{t}}{d\lambda^h}\Big|_{\bar{t}-t^*=const.}$$

$$= \frac{(p(t)+\lambda^h\frac{\partial p(t)}{\partial\lambda^h})[D+(M-D)e^{-\lambda^h(\bar{t}-t^*)}]-(\bar{t}-t^*)\lambda(\bar{t})(M-D)e^{-\lambda^h(\bar{t}-t^*)}}{\lambda'(\bar{t})[D+(M-D)e^{-\lambda^h(\bar{t}-t^*)}]-\lambda^h\lambda(\bar{t})(M-D)e^{-\lambda^h(\bar{t}-t^*)}}.$$

The sign of this last effect is ambiguous. Let me now turn to the indirect effects through a change in the delay period. The change in the length of the delay period has a direct effect given by

$$\frac{d\bar{t}}{d(\bar{t}-t^*)} =$$

$$\frac{\lambda^h\lambda(t)(M-D)e^{-\lambda^h(\bar{t}-t^*)}}{\lambda'(\bar{t})[D+(M-D)e^{-\lambda^h(\bar{t}-t^*)}]-\lambda^h\lambda(\bar{t})(M-D)e^{-\lambda^h(\bar{t}-t^*)}} < 0.$$

The total effect of a change in the parameter set (denoted by dP) is given by

$$\frac{d\bar{t}}{dP} = \frac{d\bar{t}}{dP}\Big|_{\bar{t}-t^*=const.} + \frac{d\bar{t}}{d(\bar{t}-t^*)}\frac{d(\bar{t}-t^*)}{dP},$$

where the last expression is given from the last theorem. Both the direct and indirect effect of a change in monopoly profits increases $\bar{t}$. A higher discount rate leads to an increase of $\bar{t}$ only by a positive indirect effect. Concerning the R&D-expenditure rate c there is only a negative direct effect. The direct effect from a change in duopoly profits is positive, while the indirect effect is negative. In the case of a change in the efficiency parameter the total effect is indeterminate. □

A.6 Proof of Theorem 5.1

Theorem 5.1 *An optimal incentive contract in the case with serially uncorrelated profits induces optimal effort for the leveraged firm. It can be characterized by a list of repayment obligations* $\bar{R} = (\bar{\pi}, \bar{\pi}, \pi_{00}, \pi_{01})$*, and a list of probabilities for second period finance* $\bar{\beta} = (1, 1, \frac{\pi_{00}-\pi_{01}}{\bar{\pi}-\pi_{01}}, 0)$*. In the second period the firm pays back* π_{01}*, irrespectively of which profit it realized. This contract is only offered if the expected revenue to the investor is positive, which is given by*

$$\begin{aligned} R &= \theta_i(\bar{\pi} + \pi_{01} - F) + \theta_j(1 - \theta_i)\pi_{01} \\ &\quad + (1 - \theta_i)(1 - \theta_j)(\pi_{00} + \frac{\pi_{00} - \pi_{01}}{\bar{\pi} - \pi_{01}}(\pi_{01} - F)) - F. \end{aligned}$$

Proof: The optimal repayment in the second period is derived in the text. For the first period repayment scheme and the probabilities of future finance I proceed stepwise. The Theorem is a direct consequence from the following lemmas:

First, note that only three of the (IC)-constraints are actually independent.

Lemma A.1 *The constraints (IC 2), (IC 3) and (IC 5) are redundant.*

Proof: Add the constraints (IC 1) and (IC 4) to replicate (IC 2). Add the constraints (IC 4) and (IC 6) to replicate (IC 5). Add the constraints (IC 1), (IC 4) and (IC 6) to replicate (IC 3). □

I assumed that $F > \pi_{01}$. Under this condition I have

Lemma A.2 *The optimal contract provides no second period finance, if the lowest profit was reported in the first period. Repayment is then* π_{01} *in the first period.*

Proof: If $F > \pi_{01}$, the investor always looses money in the second period. Therefore, he chooses the probabilities for future finance as low as possible, while providing sufficient incentives. If first period profits are low anyway, there is no incentive problem. Therefore the investor chooses $\beta_{01} = 0$ and the maximum repayment consistent with the (IR)-constraint of the firm, which is $R_{01} = \pi_{01}$.□

The constraint (IC 6) can thus be rewritten as

$$\text{(IC 6')} \quad R_{00} \leq \pi_{01} + \beta_{00}(\bar{\pi} - \pi_{01})$$

Let me further assume that equilibrium effort rates in the second period are sufficiently high, such that $\bar{\pi} > \pi_{00}$. If this condition holds, I have from (IC 6') and the previous lemma

$$\beta_{00} \leq \frac{R_{00} - \pi_{01}}{\bar{\pi} - \pi_{01}} \leq (\text{by (IR (F))}) \frac{\pi_{00} - \pi_{01}}{\bar{\pi} - \pi_{01}} < 1. \tag{A.2}$$

Now I can derive

Lemma A.3 *If the reported profit is π_{00}, $R_{00} = \pi_{00}$.*

Proof: Suppose (IC 6') is binding. It follows that $\beta_{00} = \frac{R_{00}-\pi_{01}}{\bar{\pi}-\pi_{01}}$. The marginal gain from lowering R_{00} (and therefore lowering β_{00}) is

$$\frac{\partial \left(\frac{R_{00}-\pi_{01}}{\bar{\pi}-\pi_{01}} (F - \pi_{01}) \right)}{\partial R_{00}} = \frac{F - \pi_{01}}{\bar{\pi} - \pi_{01}} < 1,$$

where the last inequality stems from the fact that the innovation is profitable in expected terms. The investor thus chooses the maximum repayment in this case. If (IC 6') is not binding, she chooses the maximum repayment, anyway.□

Indeed, I can state which constraints are binding.

Lemma A.4 *The incentive compatabiliy constraints (IC 1), (IC 4) and (IC 6') are binding at the optimal contract.*

Proof: I show that (IC 6') is binding. The proof for the other constraints proceeds in a similar fashion. Suppose, it were optimal to choose β_{00} such that $\beta_{00} > \frac{\pi_{00}-\pi_{01}}{\bar{\pi}-\pi_{01}}$. Since the investor loses money in the second period, he gains from lowering β_{00}. This leads to a contradiction.□

As a corollary, I have

Lemma A.5 *If profits are serially uncorrelated, optimal effort is induced by the optimal incentive contract.*

Proof: Plugging the (IC)-constraints into condition (E) gives equation (2).□

Furthermore, there is an upper limit for repayment if profits are high:

Lemma A.6 *If the highest or second-to-highest profit is reported, there will always be finance in the second period, and first-period repayment is* $\bar{\pi}$.

Proof: To show that the repayment is $\bar{\pi}$, note that

$$R_{10} - \pi_{00} = \left(1 - \frac{\pi_{00} - \pi_{01}}{\bar{\pi} - \pi_{01}}\right)(\bar{\pi} - \pi_{01})$$

from previous lemmas. Thus, I have $R_{10} = \bar{\pi}$. The same argument holds for R_{11}. The direct marginal gain of decreasing β_{10} is $\pi_{01} - F$. At the same time there is a loss because first-period repayment is decreasing in β_{10} from (IC 2). The marginal loss is $\bar{\pi} - \pi_{01}$. The net effect is therefore negative. Thus, $\beta_{10} = 1$. Again, the same argument holds for β_{11}.

Note that this contract is only offered if the participation constraint of the investor is fulfilled. This constraint is given by (IR(I)). Plugging in the optimal contract specifications yields

$$\begin{aligned} R &= \theta_i(\bar{\pi} + \pi_{01} - F) + \theta_j(1 - \theta_i)\pi_{01} \\ &\quad + (1 - \theta_i)(1 - \theta_j)\left(\pi_{00} + \frac{\pi_{00} - \pi_{01}}{\bar{\pi} - \pi_{01}}(\pi_{01} - F)\right) - F. \end{aligned} \tag{A.3}$$

This completes the proof.□

References

Acs, Z.J. and Audretsch, D.B., *Innovation and Small Firms*; Cambridge [1990].

Aghion, P. and Bolton, P., "An Incomplete Contract Approach to Financial Contracting," *Review of Economic Studies* Vol. 59 [1992], 473-94.

Arrow, K.J., "Economic Welfare and the Allocation of Resources for Invention," in R.R. Nelson: *The Rate and Direction of Inventive Activity*; Princeton [1962].

Ayres, R.U., "Toward a non-linear dynamics of technological progress," *Journal of Economic Behavior and Organization*, Vol. 24 [1994], 35-69.

Bag, P.K. and Dasgupta, S., "Strategic R&D Success Announcements," *Economics Letters*, Vol. 47 [1995], 17-26.

Berndt, E.R., Bui, L.T., Reiley, D.H. and Urban, G.L., "The Roles of Marketing, Product Quality and Price Competition in the Growth and Composition of the U.S. Anti-Ulcer Drug Industry," Sloan School of Management DP. No. 19-94, Cambridge [1994].

Bester, H., "The Role of Collateral in Credit Markets with Imperfect Information," *European Economic Review*, Vol. 31 [1987], 887-99.

Bester, H. and Hellwig, M., "Moral Hazard and Equilibrium Credit Rationing: An Overview of the Issues," in G. Bamberg and K. Spremann: *Agency Theory, Information and Incentives*; Berlin [1987].

Bhattacharya, S., Chatterjee, K. and Samuleson, L., "Sequential Research and the Adoption of Innovations," *Oxford Economic Papers (Supplement)*, [1986], 219-43.

Bhattacharya, S. and Mookherjee, D., "Portfolio Choice in Research and Development," *RAND Journal of Economics*, Vol. 17 [1986], 594-605.

Bhattacharya, S. and Thakor, A., "Contemporary Banking Theory," *Journal of Financial Intermediation*, Vol. 3 [1993], 2-50.

Bolton, P. and Farrell, J., "Decentralization, Duplication, and Delay," *Journal of Political Economy*, Vol. 98 [1990], 803-26.

Bolton, P. and Scharfstein, D.S., "A Theory of Predation Based on Agency Problems in Financial Contracting," *American Economic Review*, Vol. 80 [1990], 93-106.

Brander, J.A. and Lewis, T.R., "Oligopoly and Financial Structure: The Limited Liability Effect," *American Economic Review*, Vol. 76 [1986], 956-70.

Budd, C., Harris, C. and Vickers, J., "A Model of the Evolution of Duopoly: Does the Asymmetry Between Firms Tends to Increase or Decrease?," *Review of Economic Studies*, Vol. 60 [1993], 543-73.

Bulow, J.I., Geanakoplos, J.D. and Klemperer, P.D., "Multimarket Oligopoly: Strategic Substitutes and Complements," *Journal of Political Economy*, Vol. 93 [1985], 488-511.

Chamley, C. and Gale, D. "Information Revelation and Strategic Delay in a Model of Investment." *Econometrica* Vol. 62 [1994], 1065-86.

Choi, J.P., "Dynamic R&D Competition under "Hazard Rate" Uncertainty." *RAND Journal of Economics*, Vol. 22 [1991], 596-620.

Chow, Y., Robbins, H. and Siegmund, D., *Great Expectations: The Theory of Optimal Stopping;* Boston [1971].

Cohen, W.M. and Levin, R.C., "Empirical Studies of Innovation and Market Structure," in R. Schmalensee and R.D. Willig: *Handbook of Industrial Organization (II)*; Amsterdam [1989].

Cohen, W.M. and Levinthal, D.A., "Innovation and Learning: The Two Faces of R&D," *Economic Journal*, Vol. 99 [1989], 569-96.

Dasgupta, P. and Maskin, E. "The Simple Economics of Research Joint Ventures." *Economic Journal*, Vol . 97 [1987], 581-95.

Dasgupta, P., "The Theory of Technological Competition," in K. Binmore and P. Dasgupta: *Economic Organizations as Games*; London [1986].

Dasgupta, P., "Patents, Priority and Imitation or, The Economics of Races and Waiting Games," *Economic Journal*, Vol. 98 [1988], 66-80.

Dasgupta, P. and Stiglitz, J., "Uncertainty, Industrial Structure, and the Speed of R&D," *Bell Journal of Economics*, Vol. 11 [1980a], 1-28.

Dasgupta, P. and Stiglitz, J., "Industrial Structure and the Nature of Innovative Activity," *Economic Journal*, Vol. 90 [1980b], 266-93.

D'Aspremont, C. and Jacquemin, A., "Cooperative and Noncooperative R&D in Duopoly with Spillovers," *American Economic Review*, Vol. 78 [1988], 1133-37.

De Fraja, G. "Strategic Spillovers in Patent Races." *International Journal of Industrial Organization*, Vol. 11 [1993], 139-46.

Dewatripont, M and Tirole, J., "A Theory of Debt and Equity: Diversity of Securities and Manager-Shareholder Congruence," *Quarterly Journal of Economics*, Vol. 109 [1994], 1027-54.

Diamond, D.W., "Financial Intermediation and Delegated Monitoring," *Review of Economic Studies*, Vol. 51 [1984], 393-414.

Diamond, D.W., "Monitoring and Reputation: The Choice between Bank Loans and Directly Placed Debt," *Journal of Political Economy*, Vol. 99 [1991], 689-721.

Dixit, A. and Stiglitz, J., "Monopolistic Competition and Optimum Product Diversity," *American Economic Review*, Vol. 67 [1977], 297-308.

Dowd, K., "Optimal Financial Contracts," *Oxford Economic Papers*, Vol. 44 [1992], 672-93.

Dutta, P.K. and Rustichini, A., "A Theory of Stopping Time Games with Applications to Product Innovations and Asset Sales," *Economic Theory*, Vol. 3 [1993], 743-63.

Farrell, J., "Cheap Talk, Coordination, and Entry," *RAND Journal of Economics*, Vol. 18 [1987], 34-39.

Farrell, J. and Saloner, G., "Coordination through Committees and Markets," *RAND Journal of Economics*, Vol. 19 [1988], 235-52.

Faure-Grimaud, A., "Debt, Dividend and Imperfect Competition," *mimeo* GREMAQ, Toulouse [1995].

Fershtman, C. and Judd, K.L., "Equilibrium Incentives in Oligopoly," *American Economic Review*, Vol. 77 [1987], 927-40.

Fudenberg, D., Gilbert, R., Stiglitz, J. and Tirole, J., "Preemption, Leapfrogging and Competition in Patent Races," *European Economic Review*, Vol. 22 [1983], 3-31.

Fudenberg, D. and Tirole, J., "The Fat-Cat Effect, the Puppy-Dog Ploy and the Lean and Hungry Look," *American Economic Review (P.&P.)*, Vol. 74 [1984], 361-66.

Fudenberg, D. and Tirole, J., "A "Signal-Jamming" Theory of Predation," *RAND Journal of Economics*, Vol. 17 [1986], 366-76.

Fudenberg, D. and Tirole, J., *Game Theory*; Cambridge [1991].

Gale, D. and Hellwig, M., "Incentive-Compatible Debt Contracts: The One-Period Problem," *Review of Economic Studies*, Vol. 52 [1985], 647-63.

Gilbert, R.J. and Newberry, D.M.G., "Preemptive Patenting and the Persistence of Monopoly," *American Economic Review*, Vol. 72 [1982], 514-26.

Gilbert, R.J. and Newberry, D.M.G., "Uncertain Innovation and the Persistence of Monopoly: Comment," *American Economic Review*, Vol. 74 [1984], 238-43.

Glazer, A., "The Advantages of Being First," *American Economic Review*, Vol. 75 [1985], 73-80.

Gort, M. and Klepper, S., "Time Paths in the Diffusion of Product Innovations," *Economic Journal*, Vol. 92 [1982], 630-53.

Grossman, G.M. and Shapiro, C., "Optimal Dynamic R&D Programs," *RAND Journal of Economics*, Vol 17 [1986], 581-93.

Grossman, G.M. and Shapiro, C., "Dynamic R&D Competition," *Economic Journal*, Vol. 97 [1987], 372-87.

Harris, C. and Vickers, J., "Racing with Uncertainty," *Review of Economic Studies*, Vol. 54 [1987], 1-21.

Harris, C. and Vickers, J., "Perfect Equilibrium in a Model of a Race," *Review of Economic Studies*, Vol. 52 [1985], 193-209.

Hart, O. and Moore, J., "Default and Renegotiation: A Dynamic Model of Debt," University of Edinburgh Discussion Paper [1989].

Hart, O. and Moore, J., "A Theory of Debt Based on the Inalienability of Human Capital," *Quarterly Journal of Economics*, Vol. 109 [1994], 841-79.

Himmelberg, C.P. and Petersen, B.C., "R&D and Internal Finance: A Panel Study of Small Firms in High-Tech Industries," *Review of Economics and Statistics*, Vol. 76 [1994], 38-51.

Holmström, B. and Tirole, J, "Market Liquidity and Performance Monitoring," *Journal of Political Economy*, Vol. 101 [1993], 678-709.

Holtz-Eakin, D., Joulfrain, D. and Rosen, H.S., "Sticking it out: Entrepreneurial Survival and Liquidity Constraints," *Journal of Political Economy*, Vol. 102 [1994], 53-75.

Jensen, M. and Meckling, W., "Theory of the Firm: Managerial Behavior, Agency Costs, and Capital Structure," *Journal of Financial Economics*, Vol. 3 [1976], 305-60.

Jensen, R., "Adoption and Diffusion of an Innovation of Uncertain Profitability," *Journal of Economic Theory*, Vol. 27 [1982], 182-93.

Jewkes, J., Sawers, D. and Stillerman, R., *The Sources of Invention*; London [1958].

Kamien, M.I. and Schwartz, N.L., "Expenditure Patterns for Risky R and D Projects," *Journal of Applied Probability*, Vol. 8 [1971], 60-73.

Kamien, M.I. and Schwartz, N.L., "Timing of Innovation under Rivalry," *Econometrica*, Vol. 40 [1972], 43-60.

Kamien, M.I. and Schwartz, N.L., "On the Degree of Rivalry for Maximum Innovative Activity," *Quarterly Journal of Economics*, Vol. 90 [1976], 245-60.

Kamien, M.I. and Schwartz, N.L., *Innovation and Market Structure*; Cambridge [1982].

Katz, M., "Game-Playing Agents: Unobservable Contracts as Precommitments," *RAND Journal of Economics*, Vol. 22 [1991], 307-28.

Katz, M. and Shapiro, C., "R&D Rivalry with Licensing or Imitation," *American Economic Review*, Vol. 77 [1987], 402-20.

Kovenock, D. and Phillips, G., "Capital Structure and Product Market Behavior: An Examination of Plant Closing and Investment Decisions," Purdue University Discussion Paper No. 93-12-3 [1993].

Kovenock, D. and Phillips, G., "Capital Structure and Product Market Rivalry: How Do We Reconcile Theory and Evidence ?" *American Economic Review*, Vol. 85 [1995], *forthcoming.*

Lee, T., "On the Reswitching and Convergence Properties of R&D Rivalry," *Management Science*, Vol. 30 [1984], 186-97.

Lee, T. and Wilde, L.L., "Market Structure and Innovation: A Reformulation," *Quarterly Journal of Economics*, Vol. 94 [1980], 429-36.

Levin, R.C., Cohen, W.M. and Mowery, D.C., "R&D Appropriability, Opportunity, and Market Structure: New Evidence on Some Schumpeterian Hypotheses," *American Economic Review (P.&P.)*, Vol. 75 [1985], 20-24.

Levin, R.C., Klevorick, A.K., Nelson, R.R. and Winter, S.G., "Appropiating the Returns from Industrial R&D," *Brookings Papers on Economic Activity* [1987], 783-820.

Lippman, S.A. and McCardle, K.F., "Dropout Behavior in R&D Races with Learning," *RAND Journal of Economics*, Vol. 18 [1987], 287-95.

Lippman, S.A. and Mamer, J., "Innovation and the Persistence of Monopoly," *Economics Letters*, Vol. 38 [1992], 83-91.

Lippman, S.A. and Mamer, J., "Preemptive Innovation," *Journal of Economic Theory*, Vol. 61 [1993], 104-19.

Lippman, S.A., Mamer, J. and McCardle, K.F., "Comparative Statics in Noncooperative Games via Transfinitely Iterated Play," *Journal of Economic Theory*, Vol. 41 [1987], 288-303.

Loury, G.C., "Market Structure and Innovation," *Quarterly Journal of Economics*, Vol. 93 [1979], 395-420.

Mamer, J., "Monotone Stopping Games," *Journal of Applied Probability*, Vol. 24 [1987], 386-401.

Mansfield, E., "How Rapidly Does New Industrial Technology Leak Out," *Journal of Industrial Economics*, Vol. 34 [1985], 217-23.

Mansfield, E., Schwartz, M., and Wagner, S., "Imitation Costs and Patents: An Empirical Study," *Economic Journal*, Vol. 91 [1981], 907-18.

Mariotti, M., "Unused Innovations," *Economics Letters*, Vol. 38 [1992], 367-71.

Markham, J.W., "Market Structure, Business Conduct, and Innovation," *American Economic Review (P.&P.)*, Vol. 55 [1965], 323-32.

Martin, S., *Advanced Industrial Economics*; Oxford [1993].

McLean, R.P. and Riordan, M.H., "Industry Structure with Sequential Technology Choice," *Journal of Economic Theory*, Vol 47 [1989], 1-21.

Milgrom, P. and Roberts, J., "Limit Pricing and Entry Under Incomplete Information: An Equilibrium Analysis," *Econometrica*, Vol. 50 [1982], 443-60.

Milgrom, P. and Shannon, C., "Monotone Comparative Statics," *Econometrica*, Vol. 62 [1994], 157-80.

Mookherjee, D. and Ray, D., "On the Competetive Pressure Created by the Diffusion of Innovations," *Journal of Economic Theory*, Vol. 54 [1991], 124-47.

Myers, S.C., "Determination of Corporate Borrowing," *Journal of Financial Economics*, Vol. 4, [1977], 147-75.

Nelson, R.R., "The Economics of Invention: A Survey of the Literature," *Journal of Business*, Vol. 32 [1959], 101-27.

Opler, T.C. and Titman, S., "Financial Distress and Corporate Performance," *Journal of Finance*, Vol. 49 [1994], 1015-40.

Ordover, J.A. and Willig, R.D. "An Economic Definition of Predation: Pricing and Product Innovation." *Yale Law Journal*, Vol. 91 [1981], 8-53.

Phillips, A., "Patents, Potential Competition, and Technical Progress," *American Economic Review*, Vol. 56 [1966], 301-10.

Phillips, A., *Technology and Market Structure*; Lexington [1971].

Phillips, G., "Increased Debt and Industry Product Markets: An Empirical Analysis," *Journal of Financial Economics*, Vol. 37 [1995], 189-38.

Reinganum, J., "Dynamic Games of Innovation," *Journal of Economic Theory*, Vol. 25 [1981], 21-41.

Reinganum, J., "Nash Equilibrium Search for the Best Alternative," *Journal of Economic Theory*, Vol. 30 [1983a], 139-52.

Reinganum, J., "Uncertain Innovation and the Persistence of Monopoly," American Economic Review, Vol. 73 [1983b], 741-48.

Reinganum, J., "Uncertain Innovation and the Persistence of Monopoly: Reply," American Economic Review, Vol. 74 [1984], 243-46.

Reinganum, J., "Innovation and Industry Evolution," *Quarterly Journal of Economics*, Vol. 99 [1985], 81-99.

Reinganum, J., "The Timing of Innovation: Research, Development, and Diffusion," in R. Schmalensee and R.D. Willig: *Handbook of Industrial Organization (I)*; Amsterdam [1989].

Romer, P., "Long-Run Policy Analysis and Long-Run Growth," *Journal of Political Economy*, Vol. 99 [1986], 500-21.

Romer, P., *Dynamic Competitive Equilibria with Externalities, Increasing Returns, and Unbounded Growth.* Ph.D.-Dissertation, University of Chicago [1983].

Schumpeter, J.A., *Theorie der wirtschaftlichen Entwicklung*; Leipzig [1912].

Schumpeter, J.A., *Capitalism, Socialism, and Democracy*; New York [1942].

Shaked, A. and Sutton, J., "Relaxing Price Competition through Product Differentiation," *Review of Economic Studies*, Vol. 49 [1982], 3-13.

Shapiro, C., "Theories of Oligopoly Behavior," in R. Schmalensee and R.D. Willig: *Handbook of Industrial Organization (I)*; Amsterdam [1989].

Spatt, C.S. and Sterbenz, F.P. "Learning, Preemption, and the Degree of Rivalry." *RAND Journal of Economics*, Vol. 16 [1985], p 84-92.

Spence, M., "Investment Strategy and Growth in a New Market," *Bell Journal of Economics*, Vol. 10 [1979], 1-19.

Spence, M., "Cost Reduction, Competition and Industry Performance," *Econometrica*, Vol. 52 [1984], 101-21.

Tarski, A. "A Lattice-Theoretical Fixpoint Theorem and its Applications," *Pacific Journal of Mathematics*, Vol. 5 [1955], 285-309.

Tirole, J., *The Theory of Industrial Organization*; Cambridge [1988].

Topkis, D.M., "Minimizing a Submodular Function on a Lattice," *Operations Research*, Vol. 26 [1978], 305-21.

Topkis, D.M., "Equilibrium Points in Nonzero-Sum N-Person Submodular Games," *SIAM Journal of Control and Optimization*, Vol. 17 [1979], 773-787.

Townsend, R.M., "Optimal Contracts and Competitive Markets with Costly State Verification," *Journal of Economic Theory*, Vol. 21 [1979], 265-93.

v.Damme, E., "Banking: A Survey of Recent Microeconomic Theory," *Oxford Review of Economic Policy*, Vol. 10 [1994], 14-33.

Veblen, R., *Imperial Germany and the Industrial Revolution*; New York [1915].

Vickers, J., "The Evolution of Market Structure when there is a Sequence of Innovations," *Journal of Industrial Economics*, Vol. 35 [1986], 1-12.

Vives, X., "Nash Equilibrium with Strategic Complementarities," *Journal of Mathematical Economics*, Vol. 19 [1990], 305-21.

Völker, R., *Innovationsentscheidungen und Marktstruktur*; Heidelberg [1990].